Academic Entrepreneurship, Scientists, and Scientific Careers

This series addresses the three interrelated themes of academic entrepreneurship, scientists and scientific careers. The series covers all aspects of academic entrepreneurship, with respect to interpretation, engagement, implementation and outcomes of academic entrepreneurship and the adoption of core attributes and characteristics of academic entrepreneurship in publicly funded research laboratories. In particular, the editors seek perspectives from business and social science disciplines that challenge convention thinking, and welcome monographs that employ a variety of methodological approaches that seek to analyze and reimagine academic entrepreneurship. To complement and extend the existing literature, we are particularly interested in contributions that focus on emerging and transitional economy contexts. Within academic entrepreneurship, the role of the scientists and, more broadly, the issue of scientific careers, transitions and identity change has been somewhat neglected. Therefore, additional aims of this series are to stimulate and gather contributions that specifically focus on scientists and the development of scientific careers in universities, public laboratories and private sector contexts. Our scope is both on the micro and meso levels so as to illuminate the behaviors, activities and challenges faced by scientists irrespective of career stage. Of particular interest are contributions that focus on gender issues, how scientists strategize with respect to research impact, and their engagement with academic entrepreneurship.

Morgan Boyce

Principal Investigators and R&D Failure

Probability of Innovation Failure in Small Business

Springer

Morgan Boyce
Charlotte, NC, USA

ISSN 2523-7004 ISSN 2523-7012 (electronic)
Academic Entrepreneurship, Scientists, and Scientific Careers
ISBN 978-3-031-43607-9 ISBN 978-3-031-43608-6 (eBook)
https://doi.org/10.1007/978-3-031-43608-6

This Springer imprint is published by the registered company Springer Nature Switzerland AG
The registered company address is: Gewerbestrasse 11, 6330 Cham, Switzerland

To Brooklyn and Brielle
May all your failures be successes

Preface

There is a small but growing body of literature that examines the role of principal investigators in publicly funded R&D projects. This work provides a discussion of the literature on principal investigators and R&D project failure and found there is limited intersection between the two. I provide a theoretical model in this dissertation which explains how firm characteristics, including those of the principal investigator, impact the probability of failure. The theoretical model serves as a structural form model to motivate the empirical analysis which assesses the probability of failure in small technology-based firms that received a Phase II award from the Department of Energy's Small Business Innovation Research program. Using a Probit model, I estimate a reduced form specification of the structural model to estimate the probability a firm will experience failure conditional on characteristics of the principal investigator and the firm. I found that prior experience of the firm with a similar technology, university faculty involvement, and the age of the principal investigator are negatively associated with project failure. I also found a positive relationship between failure and firms where the principal investigator was the sole founder and CEO of the firm.

Charlotte, NC, USA Morgan Boyce

Acknowledgments

This book represents the culmination of many years of schooling of which several failures have contributed to the successful publication of this book. While this work likely marks the capstone of my formal education, I also hope it is the first of many such studies. To that end, I have learned from and remain thankful to several scholars that have lent an ear or contributed to the ideas of this piece either directly or indirectly.

I'd like to give a special thanks of gratitude to Dr. Albert Link for the many hours spent discussing ideas, the many reviews, the suggestions, and his general guidance and advice for the overall writing of this book. Additionally, it would be remiss not to thank Dr. Link for his witty one-liners that often accompanied his review commentary. I'd like to give a big thanks to Dr. Martijn van Hasselt for all his valuable reviews, ideas, and exceptional feedback especially related to the econometric analysis presented in this piece. I thank Dr. Dennis Leyden, for his expert guidance and thoughtful review, especially in the development of the theoretical model. I'd also like to thank Dr. James Cunningham for his detailed review, helpful comments, and guidance in navigating the proposal for this monograph. It has been a sincere pleasure and honor to have worked with each of these outstanding scholars. Thank you.

Finally, I must give a huge thanks to my wife, Alexis. From holding-down-the-fort with our two young daughters to give me the time needed to write this monograph to staying up late to give feedback on grammar and style, and the occasional motivational speech, this monograph was made possible by her. Thank you for all your love and support.

Contents

Chapter 1
Introduction

Abstract This introductory chapter begins with a brief explanation of why studying innovation and entrepreneurship is important at a macrolevel. It discusses how the landscape and expectations of public support of innovative projects are changing and also defines what innovation is in the context of this analysis. This chapter explains the importance of this area of study, and that there is limited research that examines factors associated with the failure of innovative projects at a microlevel. A description of the ways this monograph contributes to this area of research is provided followed by brief descriptions of the following chapters.

Research and development (R&D) is critical in the creation of new technology, which in turn drives innovation. Innovation is fundamental to long-term economic growth, and investments in R&D are key to the perpetuation of innovation. In a National Bureau of Economic Research (NBER) working paper, Bryan and Williams (2021, p. 6) describe the importance of innovation stating, "Unlike many "standard" goods, innovations play a cumulative and fundamental dynamic role in economic growth. When innovation policy is suboptimal, growth is as well." Feldman, Hadjimichael, Lanahan, and Kemeny (2016, p. 1) echoed this same idea stating, "… there is widespread recognition that the long-term growth of industrial economies relies on innovation, entrepreneurship, and production …" Innovation, when defined synonymously as invention, may be defined as the creation of a new idea, device, or method. Given this narrow definition, the mere creation of an innovation does not necessitate the propagation of the innovation through society. For example, an individual or firm may engage in R&D and invent a viable new device but fails to bring their innovation to market. If that innovation never becomes a commercialized product, available for consumption by society, then that new product will not contribute to economic growth.

Since performing R&D has costs, there must be capital to fund the engagement in R&D. Generally, when an investment is made in a risky venture, such as funding an R&D project, an expected return commensurate with the level of risk is required to make the investment rational. When private investments are made in R&D, the

M. Boyce, *Principal Investigators and R&D Failure*, Academic Entrepreneurship, Scientists, and Scientific Careers, https://doi.org/10.1007/978-3-031-43608-6_1

return expected is generally some form of monetary compensation. However, when public funds are used to invest in an R&D project, the return expected from that investment may come in other forms such as increased employment or some other benefit to society. If public funds are used to support a firm's R&D of an innovation/ invention, then if that innovation never makes it to market, society may never realize a positive return on its investment.

Therefore, if a publicly funded R&D project results in an innovation, there must be a mechanism to make the innovation available to society. This is where an entrepreneur plays an important role. Before continuing, we must acknowledge that entrepreneurship; hence, the entrepreneur has a significant body of literature that dates back centuries, and within this literature there have been many definitions of what an entrepreneur is.[1] Similar to the many definitions of an entrepreneur, there are many interpretations of how to define innovation. In a recent study, Singh and Aggarwal (2022) analyzed 208 definitions of innovation in search of a consensus. Table 1.1 presents a summary of the prominent definitions discussed by Singh & Aggarwal. Largely common across the way innovation is defined is the idea or concept of something that is new.

Often innovation and entrepreneurship are used synonymously, but they can be distinct concepts. There can be situations where the incentive for developing an innovation does not align with the typical incentive for an entrepreneur. While either can occur independently, innovation and entrepreneurship can also behave as a symbiotic relationship. This is akin to Joseph Schumpeter's view of an entrepreneur as an innovator as described in his theory of economic growth famously known as "creative destruction." In this book, Schumpeter's view of an entrepreneur is accepted but through a slightly different lens; that is innovator as entrepreneur. Further, the definition of innovation that most closely relates to innovation discussed in the remainder of this piece is that of Gordon and McCann (2005), bringing new product, service, or process leading to commercial success (see Table 1.1).

Policymakers prefer the symbiotic relationship between innovation and entrepreneurship when allocating public funds to R&D projects, since an innovation brought to market has a greater opportunity to increase social benefit than an innovation that remains in a lab or a research paper. An innovation brought to market also increases the return on public investment, which continues to be important to policymakers to continue to support public programs that invest in R&D projects. As stated by Link and Wright (2015, p. 1):

> Policy makers reemphasized the importance of program evaluation requirements—originally set forth in the Government Performance and Results Act (GPRA) of 1993 (P.L. 103-62)—during their prolonged deliberations (from 2008 through 2011) of the reauthorization of the publicly funded U.S. Small Business Innovation Research (SBIR[2]).

[1] Leyden and Link (2015, p. 4–13) provides a thorough discussion of the evolution of the literature spanning multiple disciplines and the various definitions of an entrepreneur.

[2] The SBIR program is relevant to this book as the data used in the empirical analysis relates to the SBIR program.

Table 1.1 Definitions of innovation

Definition of innovation	References
To introduce something new.	Aronson, Jeffrey K. 2008. "Something new every day." Journal of Ambulatory Care Management 31 (1): 65–68
An idea, practice, or object perceived as new	Rogers, E. M. 1995. Diffusion of innovations. Fourth ed. New York: The Free Press
Ability to generate, assimilate, implement use, and exploit new ideas	Cohen, W. M., and D. A. Levinthal. 1989. "Innovation and learning: The two faces of R and D." The Economic Journal 99 (397): 569–596. Damanpour, F., and S. Gopalakrishnan. 2001. "The dynamics of the adoption of product and process innovations in organizations." Journal of Management Studies 38 (1): 45–65. Hivner, W., and W. E. Hopkins. 2003. "Facilitating, accelerating, and sustaining the innovation diffusion process: An epidemic modeling approach." Journal of Innovation Management 6 (2): 80–89. Urabe, K. 1988. "Innovation and Teh Japanese Management system." In Innovation and Management: International Comparisons, edited by Kuniyoshi Urabe, John Child, and Tadao Kagono, 3–7. New York: Walter de Gruyter
Creation of a new combination of resources	Schumpeter, J. A. 1952. Theorie Der Wirtschaftlichen Entwicklung. 5th ed. Berlin: Duncker & Humblot
Creation of new opportunities by harnessing creative ideas	Zhang, Wen. 2011. "Research on Innovation-Generating and Innovation-Transferring." In 2011 International Conference on Management and Service Science, 1–4. IEEE.
Creation of tangible social value through fresh thinking or by introducing something new	Velo, G., and P. Rizzini. 2008. "Erice statement on drug innovation." British Journal of Clinical Pharmacology 65 (3): 440–441
The process of creation of market-focused value that significantly impacts the environment	Drucker, Peter. 1973. Management: Tasks, Responsibilities, Practices. New York: Harper & Row Publishers
Leading to the commercialization of the invention	Gaynor, G. H. 2013. "Innovation: Top down or bottom up." IEEE Engineering Management Review 41 (3): 5–6.
Development and operationalization of a valuable technology	Afuah, A. 2003. Innovation Management. New York: Oxford University Press
A disruptive force that creates new market leaders	Christensen, C. M. 1998. "The evolution of innovation." In The Technology Management Handbook, edited by R. C. Dorf, 3.2–3.10. Berlin: CRC press.
Bringing new product, service, or process leading to commercial success	Gordon, I. R., and P. McCann. 2005. "Innovation, agglomeration, and regional development." Journal of Economic Geography 5 (5): 523–543

Public support of innovation has occurred in the United States for over a century, but public support of innovation with the goal of increasing social benefit by bringing innovation to market has gained more attention in recent decades. In fact, there has been a notable shift toward an expectation of commercialization of the output from publicly supported (i.e., publicly financed) R&D activities, especially in universities. Boehm and Hogan (2014, p. 134) discuss this point stating, "The commercialization of scientific knowledge has become a primary objective for universities worldwide."

In the remainder of this book, innovation is defined as a new technology put into use or more specifically technology that enters the market as a product, process, or services. This definition of innovation encompasses both the R&D that goes into creating a new product, process, or service and the entrepreneurial activities to bring that innovation to market, i.e., innovator as entrepreneur. This definition of an innovator coincides with the Schumpterian view of an innovator as someone who commercializes a new idea. Further, it is not uncommon for a firm to have the goal of developing a new product with the intention of bringing it to market to earn a profit. In fact, commercialization of technology is becoming a traditional measure of R&D success, especially when the R&D is publicly supported. Although the commercialization of publicly funded R&D is becoming a desirable outcome, there is limited research related to the characteristics of key individuals responsible for R&D and their association with commercialization success or failure.

Understanding factors that are associated with publicly funded R&D project success/failure is important, and the literature examining this topic is small but growing. It is widely accepted that technological change is the driver of long run economic growth and technological change is a result of R&D (Link & Wright, 2015). Therefore, understanding drivers of R&D project failure may create the knowledge to help reduce R&D project failure, thereby, through technological change, increasing the prospects for long-run economic growth. In a study examining project failure at the National Institute of Health (NIH), Andersen et al. (2017, p. 432) echo this sentiment stating, "Identifying covariates with project failure might provide the NIH, as well as other public agencies that competitively fund firms' research, targets to increase the effectiveness of its research awards." Increasing the effectiveness of awarding public funds for R&D can help reduce publicly funded R&D failure and therefore increase the public's return on its investment. Therefore, reducing R&D project failure has economic importance due to the potential to contribute to long-run economic growth and by increasing return on the public's investment.

Principal investigators (PIs) are critical in the outcome of innovative R&D projects; therefore, their success may have an impact on long-run economic growth through the advancement of technological change. In a study aimed at understanding what factors drive scientists to becoming publicly funded principal investigators, Cunningham et al. (2016b, p. 779) state that, "… publicly funded principal investigators are the linchpin of knowledge transformation …" Understanding the association between PIs and R&D failure not only provides insights that can have a downstream impact to economic growth, it can also provide policymakers with additional information to aid decisions in allocating funds for R&D. It is known that PIs are critical in delivering successful outcomes of publicly funded innovative R&D projects due to the role they play. Cunningham et al. (2016c, p. 386) summarizes the importance of PIs in publicly funded projects noting that:

> PIs are responsible for all aspects of publicly funded projects that they lead. They manage the research programme, budgets, people, liaise with internal university functions such as finance, human resources and technology transfer as well collaborating with other academic, industry and policy partners in delivery of these projects.

Relative to the importance of PIs to the outcomes of publicly funded R&D projects, the literature examining the role PIs play is small, however, is growing. Further, there is a notable absence of literature that examines principal investigators' role in the success or failure of R&D projects; in this monograph, these issues are addressed.

In addition to contributing to the literature through a thorough review of R&D project failure and PIs, this monograph provides a new-to-the-literature theoretical model that explains how firm characteristics, including those of the principal investigator, impact the probability of failure. The theoretical model serves as a structural form model to motivate the empirical analysis that assesses the probability of failure in small technology-based firms that received a Phase II award from the US Department of Energy's Small Business Innovation Research program, which is the largest publicly funded program in the United States for small business innovation. The data for the empirical analysis has yet to be described in the literature. Using a probit model, a reduced form specification of the structural model is used to estimate the probability a firm will experience failure conditional on characteristics of the principal investigator and the firm. These specifications introduce PIs and their associated characteristics as new-to-the-literature covariates with R&D project failure.

The remainder of this book is organized as follows: Chap. 2 reviews the relevant literature related to R&D project failure. This multidisciplinary literature is limited, and there is a very limited set of literature that considers how characteristics of a key individual, such as PI, in the R&D process are associated with project success or failure. This chapter also provides a review of the literature on PIs. There is a notable lack of intersection between the literature on R&D project failure and PIs. Perhaps an explanation for the limited set of literature on PI's association with R&D project failure is the lack of data to support empirical research on research failure.

The history of DOE is discussed in Chap. 3, which shows the relevance of public/private collaboration in innovation that led to the increased well-being of society. It also provides a backdrop for the data used in the empirical analysis presented in this book. The DOE data used are related to its Small Business Innovation Research (SBIR) program.

Chapter 4 overviews the legislative history of the SBIR program along with DOE's historical involvement in the program. Also, Chap. 3 provides a review of the relevant economics literature related to the SBIR program. One conclusion from the literature review is that there is a body of research that identifies covariates with the programmatic success of an SBIR project when measured in terms of the likelihood of project commercialization. Absent from that literature is a detailed theoretical analysis of SBIR project failure, project failure being the focus of this book.

Chapter 5 presents a new-to-the-literature microeconomic theoretical model that describes how firm or project characteristics may be associated with the success or failure of a research project. The model is cast in terms of the success or failure of the search process for a technology that can be commercialized. The theoretical model introduced in this chapter is used to inform the empirical analysis presented in Chaps. 7 and 8.

A discussion of the DOE SBIR data used to examine the testable hypotheses from the theoretical model and descriptive statistics on the dependent variables from the DOE SBIR data are provided in Chap. 6. This chapter provides the first known analysis of the specific DOE SBIR data used in this book.

Chapter 7 provides findings from the initial empirical analysis that focuses on firm and project characteristics and their association with R&D project failure. A reduced form specification of the theoretical model introduced in Chap. 5 is estimated using a probit model and provides support for two previous studies' findings related to project failure.

Chapter 8 introduces PI-specific factors to the reduced form specification discussed in Chap. 7. The PI variables augment the firm and project characteristic findings examined in Chap. 7.

Chapter 9 concludes this book with a summary of the empirical findings, policy recommendations, and a suggested roadmap for future research.

Chapter 2
R&D Project Failure and Principal Investigators

Abstract This chapter reviews the literature on research and development (R&D) project failure, which spans both the economics and management literature. The literature shows that failure is often defined in terms of the lack of success of an undertaking, and there may be varying degrees of failure coupled with varying degrees of success. Failure can also be thought of as a binary descriptor to describe the situation when the initial undertaking was not fully successful in terms of the original plan. No matter how failure is defined, the literature examining how key individuals involved in a project are associated with project failure is sparse, especially that of R&D project failure. A review of the literature on principal investigators (PIs) is also provided in this chapter. There is a small but growing body of literature related to PIs where much of the existing literature is aimed at understanding the role PIs play in the R&D process. The literature conveys that PIs are critical participants in the R&D process and often have multifaceted roles within small entrepreneurial firms. An overall conclusion from the review of project failure and PIs is that there is a conspicuous absence of studies that focused on research success or failure and PI association with these outcomes.

2.1 R&D Project Failure

Failure is often defined in terms of the lack of success of an undertaking. By this definition, the definition of failure is not independent of the definition of success. There may be varying degrees of failure coupled with varying degrees of success. This situation may arise in many ways during a research and development (R&D) project, especially if the undertaking is able to be measured for completeness at distinct intervals or by independent project components/goals. However, failure can also be thought of as a binary descriptor to describe the situation when the initial undertaking was not fully successful in terms of the original plan. Binary measures of failure show up frequently in various pursuits and can easily be identified, for example, pass/fail, win/lose, accepted/rejected, sale/no sale, and the list goes on. As

M. Boyce, *Principal Investigators and R&D Failure*, Academic Entrepreneurship, Scientists, and Scientific Careers, https://doi.org/10.1007/978-3-031-43608-6_2

such, measures of success or failure can be thought of in multiple ways, and often the data available to analyze these results determine how the success or failure metric is defined. In the remainder of this monograph, unless otherwise noted, failure refers to the general case of not succeeding, and R&D failure or R&D project failure may be used synonymously. Additionally, R&D failure or R&D project failure should be considered as a specific type of the more general case of project failure.

2.1.1 Defining Success and Failure

The metrics for success related to R&D projects that have previously been studied have generally focused on the purposeful output or outcome of a business or project, and from an empirical perspective, success has been modeled as a function of project and firm characteristics. However, only a few researchers have considered how key individuals within the firm contributed to success.

As an example of a study of how key individuals contributed to success, Mansfield and Wagner (1975) collected detailed data on three firms in the industrial R&D sector and defined three measures of the probability of success. They classified the projects in their sample as either technology-push or demand-pull projects[1]; they found that technology-push projects are likely riskier and less likely to succeed than demand-pull projects. Using semi-structured interviews of key personnel collected from a two-phase sampling of 103 projects from six firms from 1969 to 1973, Rubenstein et al. (1976) defined several measures of success. The study elaborates that measures of success are specific to each project, and factors associated with success encompass human, social, and communication factors. The authors stated that success cannot be achieved by the organization as a whole, but in an overwhelming number of cases, a key individual was critical to the project's success.

Siegel and Wessner (2012) used data from the National Research Council (NRC) database of Phase II Small Business Innovation Research (SBIR)-funded projects to define seven output measures of success and study factors associated with each measure. Two of their defined measures of success were expected sales and actual sales, which were shown to be positively related to the age of the project, entrepreneurial experience of the founder, and size of the award. In a study looking to understand how behaviors of key individuals, namely project managers, shape the success of a project, Nixon et al. (2012, p. 210) found that "… research, leadership style and personal traits have also been identified as a critical success factor, determining either the success or failure of a project."

[1] Technology-push projects typically do not have a current market demand, whereas demand-pull projects are typically initiated in response to a current market demand.

2.1.2 *Measuring Business Failure*

Why Businesses Fail

As mentioned above, understanding factors associated with a particular key individual involved with the success of a project or small business has been recognized by only a few researchers, thus is an area of research that should be further developed. Similarly, understanding how individuals may contribute to the failure of a project, in particular an R&D project, should be further developed, and such studies are even more limited than studies of project success.

Project failure could be thought of as a perfect complement to success; however, that logic may not always hold. Using survey data from 97 project managers to explain why projects fail, Pinto and Manuel (1990) suggested that failure is not a binary outcome; it can vary across projects as well as in various phases of the project. They suggested that there are three distinct benchmarks to measure project performance: the implementation process itself, the perceived value of the project, and client satisfaction with the delivered project.

Shepherd and Wiklund (2006) reviewed literature aimed at answering why businesses fail. They discussed that failure is hard to define, should be studied as a process as opposed to an instantaneous event, and that economic reasons should define a business failure. They defined three causes of business failure: liability of newness, overconfidence, and lack of human capital including experience. In this sense, failure is defined as a binary outcome as opposed to a more fluid measure.

Liability of newness can be associated with the business itself. Gicheva and Link (2016) consider such liability in a study designed to estimate the probability of an SBIR project resulting in commercialization conditional on not failing. They used data from the NRC database on Phase II–funded SBIR projects to derive their estimates and found that nascent firms were 20 percent more likely to fail than a non-nascent firm.

Overconfidence and lack of human capital are factors that link human characteristics to failure but not necessarily a specific individual trait, given groups of people, such as a management team, can exert similar traits. For example, using data on 18 R&D projects that failed from the Israeli bio-medics electronic sector, Spiller and Teubal (1993) examined behaviors that are associated with firms' failed R&D projects. They found that project failure can be brought on by inappropriate firm behavior or because of uncertainty or both. Their study elaborated that choice of program and faulty program execution are examples of inappropriate firm behavior, both of which are likely undertaken by some form of management or project team. Similarly, Sauser et al. (2009) analyzed information from the National Aeronautics and Space Administration's (NASA's) Mars Climate Orbiter project with a particular focus on managerial decisions' impact on project failure. They found that in many cases, it is not technical details, mishaps, or poor designs that lead to project failure; it often results from managerial ineffectiveness.

As suggested by Rubenstein et al. (1976), a key individual may possess certain characteristics that ultimately determine the failure or success of a project. Link and Wright (2015) conducted a study to understand what drivers lead to SBIR project failure. One of their explanatory variables was the gender of the principal investigator (PI) assigned to the project. They found that if the principal investigator was a woman, there was a reduction in the probability of project failure. One explanation of this result may be that women have inherent characteristics, such as higher risk aversion compared to their male counterparts, which results in them failing less often on average. In a similar study, Andersen et al. (2017) considered why SBIR projects fail. They used data from a sample of 461 projects funded by the NIH SBIR program and found gender to be a significant factor, too. Specifically, they found that if the founder of the small business was a woman there was a negative correlation with project failure. They also found that if the business founders had a background in business there was negative correlation with project failure.

The literature around project failure has often suggested that the reasoning for failure cannot be defined universally across all projects. Failure can be a perfect complement to success, or it can be accompanied by partial success. However, the majority of the studies related to project failure have examined the reasons for project failure at the firm, managerial, or project level and have not explicitly focused on more microlevel dynamics within each project. Although each project may be unique in many facets, common across the majority of publicly funded R&D projects is the presence of a principal investigator. The PI is someone who has the overall responsibility for the project and therefore must have a level of education and experience that deems them credible for leading the project (SBIR-STTR, 2018b). For this reason, it follows that an understanding of the role PIs play in projects and the various factors associated with them may provide further insight into understanding the failure of projects. The literature on PIs is relatively small, though growing as the importance of PIs is becoming known; the performance of PIs may provide insight into understanding how likely it is that a project will fail. In the next section, the literature on PIs is reviewed, which confirms there is a lack of research that examines key individuals who are involved in R&D and their association with project failure.

2.2 Principal Investigators

PIs play a critical role in the scientific community as lead scientists, innovators, and more recently as effective managers. They are the leaders of R&D projects and, therefore, are critical individuals in the R&D process. It is important to deepen our understanding of PI's involvement in the innovation process since they are the people leading R&D projects that can bring new technologies to market that provide a large social benefit. Understanding PI's role in publicly funded R&D projects should shed light on how their characteristics are associated with project failure. The remainder of this chapter reviews the relatively small but growing body of literature on PIs, which has a central focus on the importance and multifaceted responsibilities of PIs.

2.2.1 Multifaceted Responsibilities

In a recent report by the Joint Research Centre (JRC), the authors highlight the dichotomous role of PIs being scientists and managers stating, "The PC [PI] has primary responsibility for creation of the project concept in almost 60% of projects surveyed. The PC [PI] also has the primary responsibility for selection of project partners and planning timelines and budgets in a majority of projects" (Cunningham et al., 2020, p. 3). The behavioral importance of PIs, especially publicly funded PIs, has been growing as the institutional landscape evolves. Cunningham et al. (2016a), using survey data based on semi-structured interviews of the population of publicly funded PIs in Ireland over the years 2009–2014, examined the roles and activities of the scientists who are publicly funded principal investigators. They found that the roles of the principal investigator include being a project manager, an administrator, a science broker, and more recently a boundary spanner (i.e., ability to bridge different areas and domains such as the academic sector and the private sector).

For publicly funded PIs acting as scientists, it is standard practice to share their research outcomes through traditional channels such as scientific papers and conference presentations (Baglieri & Lorenzoni, 2014). However, PIs now face additional requirements outside those a traditional scientist would encounter. Public programs, especially those that are technology-focused, also expect the PI or the PI's funded organization to commercialize from their research. This paradigm has augmented the role and responsibilities of the traditional PI. Cunningham et al. (2016a, p. 67) discuss the growing importance of PIs stating that, "publicly funded PIs are key assets and the combination of their novel efforts and their capability to meet the expanding PI role means they are a core and critical actor in transforming scientific, economic and societal environments." In this sense, PIs are innovators who are facing increasing demand to become entrepreneurs.

PIs have generally been referred to as *knowledge brokers* (e.g., Kidwell, 2013), and they are key actors in bringing their innovations to market. Using data from a survey of 135 universities, Thursby et al. (2001) found that 71 percent of early-stage inventions required the inventor's cooperation to successfully commercialize their invention. They found that industry partners required the specialized knowledge only the inventor possessed to be able to move the innovation from early-stage development to a product able to be commercialized. With reference to Mangematin et al. (2014) and Cunningham et al. (2016b), the latter (p. 779) sum up the importance of publicly funded PIs, stating that "... publicly funded principal investigators are the linchpin of knowledge transformation through articulation of research programmes, the shaping of research avenues and the bridging of academia and industry." Further, in a paper that examined the social origins of innovation failure, Pedraza (2017, p. 441) discussed the importance of building teams stating, "... bridging large cognitive distances—often a prerequisite for breakthrough innovation—requires the frequent, face-to-face interaction of members from the relevant distant communities." While the author does not specifically discuss PIs, as discussed previously, PIs are the team members who carry this responsibility. Therefore,

Pedraza's finding further supports the notion that PIs are critical participants in the innovation process.

Because PIs are the scientists who generate the ideas for innovations and are in large part responsible for the success of their R&D projects, public funding agencies are de facto choosing to fund specific PIs. In fact, according to the SBIR program:

> Every SBIR … proposal must designate a single individual who will serve as the principal investigator on the proposed project. The PI has overall responsibility for the project, and therefore must be credible in terms of his/her education, work and project management experience. (SBIR-STTR, 2018b, p. 1)

Having overall responsibility of a research project requires the PI to oversee the day-to-day operations of the project, provide updates to stakeholders, manage staff, sign off on budgets and financial plans, ensure deadlines are met, and submit technical documentation (Cunningham et al., 2016a). When a funding agency allocates resources to an R&D project, it is de facto trusting the PI to accomplish the mission of the project successfully.

The designation of PI may be viewed as an accomplishment that recognizes a scientist as reaching a certain level or milestone in their career. Being a PI conveys status within the academic community and with that accreditation, additional resources may be allocated to the PI to assist in endeavors that add both economic and noneconomic value to the research (Cunningham et al., 2017). Because becoming a PI does not happen by chance, and given their high rank in the scientific community, it is useful to understand how or why a person is elevated to PI. In a case study of 30 projects in Ireland's science, engineering, and technology sector from 2009 to 2014, Cunningham et al. (2016b) found that scientists tended to become PIs because of either a pull factor or a push factor. The three pull factors identified were control over the project, motivation or career advancement, and ambition/personal drive. The two push factors identified were a lack of other options or project dependencies and institutional pressures.

No matter the reason or reasons that a person becomes or is designated as a PI, being a PI results in additional responsibility beyond that of a research scientist. As stated above, the PI has the overall responsibility for an R&D project and therefore must be multifaceted in their abilities. Using survey data based on semi-structured interviews of the population of publicly funded PIs in Ireland over the years 2009–2014, Cunningham et al. (2016a) found that PIs must act as project managers, administrators, and science brokers and have the ability to bridge the gap between multiple domains such as the private and academic sectors. Kidwell (2013) echoed the notion of PIs being boundary spanners through a case study of four PIs, involved in nanotechnology research funded by the SBIR program, who have successfully commercialized their research. Similarly, Boehm and Hogan (2014) collected data from semi-structured interviews of 82 stakeholders involved in 17 collaborative research projects in German and Irish universities, 25 of the interviews were with PIs. Using these data, they found that PIs must take on multiple roles such as project managers, negotiators, resource acquirers as well as traditional academics.

2.2.2 The Goal of Commercialization

PIs, as boundary spanners, play an important role in commercialization-related activities associated with R&D projects. PIs are at the center of the entrepreneurial network system that includes funding agencies, venture capitalists, banks, entrepreneurs, and high-growth small and medium enterprises (SMEs) (Cunningham et al., 2017). Each participant in the entrepreneurial process seeks to maximize their return on investment, albeit a monetary or nonmonetary investment. The PI's role becomes even more challenging when participants in the process have competing interests. Cunningham et al. (2014) discussed the inhibiting factors associated with PIs leading publicly funded research programs. They found a key tension between entrepreneurial and scientific outcomes. The PI's employer and other stakeholders may pressure the PI to commercialize their research, but as a scientist, the PI may be less interested in capitalizing on the research and more interested in moving the science forward (these objectives may not be mutually exclusive).

Commercialization of the firm's output requires the PI to engage in entrepreneurial activities. The push by the PI's employer and other stakeholders to commercialize their research does not necessarily translate to the PI having a strong desire to be a successful entrepreneur. PIs, as project leads, have some autonomy when it comes to how they allocate their time across the various boundary-spanning tasks. Del et al. (2017) used data from the Eurobarometer survey and a questionnaire to PIs in Europe to understand what factors are associated with PI's preferences toward entrepreneurial activities and their performance. They found that PIs still lean toward research over entrepreneurial activities, but that a country's culture toward entrepreneurship may influence a PI to have a greater preference toward entrepreneurial activities. PIs may prefer research activities over entrepreneurial activities; however, the ability to acquire external funding is increasingly being viewed as a core competency (Geuna & Nesta, 2006). Further, since PIs are at the center of the entrepreneurial ecosystem, they are by default key actors in transferring knowledge and commercialization activities (Cunningham et al., 2016a, b, c; Menter, 2016; O'Kane, 2016). For these reasons, commercialization is an activity growing in importance that PIs may have to embrace.

When forming a research proposal, the PI must have an agenda for the project including whether to attempt to commercialize the research. With any innovative work, there will be risk, uncertainty, or both associated with achieving the project's goals. The presence of either risk or uncertainty should not alter a project's goals or deter the project itself from being undertaken as long as the risk or uncertainty is not prohibitive. O'Kane et al. (2017) discuss particular risk factors associated with commercialization of publicly funded R&D projects. Using data from semi-structured interviews with 24 funded health science PIs in New Zealand, the authors found four factors that inhibit PIs from incorporating or completing commercialization as a part of their research plan. The factors are (2017, p. 216):

> (1) PI-funding body trust; (2) disconnects between universities and funding bodies' expectations; (3) deficiencies in TTO [technology transfer office] resourcing; and (4) levels of conflict/complementarity between publication and commercialization activities in funded science.

Using ten case studies of PIs from Irish universities, O'Reilly and Cunningham (2017) examined the barriers and enablers to successful technology transfer to SMEs from university research and the role the university plays in the success. They found asset scarcity as a barrier to successful technology transfer with SMEs. Also considered was the geographic proximity of the SME to the PI, but this was not referenced as a barrier to commercialization. However, social and cultural proximities were deemed significant. Further, an SME's perceptions of bureaucracy within university technology transfer offices (TTOs) were cited as a barrier, and in some cases, SME research relationships were undervalued because of the expectation of low financial returns.

When a PI oversees an R&D project that has received outside funding, the funding party expects the PI to deliver a successful project. If a PI fails to deliver a successful project, the funding party may begin losing trust in the PI's capabilities. The commercialization success of a research project has been studied as a barometer for measuring PI's successfulness. Recognizing success can have many different faces, Del et al. (2017) identified four key performance indicators (KPIs) to measure PI's entrepreneurial intentions and performance in a cross-country comparison. The KPIs are networking and resources acquired, innovations realized, technology transfer activities, and new spin-offs and start-ups. Kidwell (2013) analyzed the characteristics associated with PIs involved in nanotechnology who successfully commercialized their research. Kidwell found that prior success in commercializing their research has a strong correlation with future success. Further, successful PIs build trust with stakeholders by presenting solutions to a problem even when the industry is not initially aware of the problem. They build trust by anticipating the future and managing issues or conflicts. Essentially, successful PIs find potential industry partners and build a good relationship with them. Further, highlighting the importance of PI's reputation or track record of success, Boehm and Hogan (2014) found that industry partners may favor a specific PI regardless of the university that employs them.

2.3 Conclusion

Firm or project failure has long been studied, but there is a notable deficit in studies that examine how key individuals in the firm/project team are associated with firm or project failure. There is an even smaller body of literature that examines factors associated with R&D project failure and how a key individual, such as a PI, may be associated with the R&D project outcome. In general, the studies that have analyzed factors associated with R&D project failure have considered overall firm or market characteristics that have contributed to the outcome of the R&D project. Further developing the literature on R&D project failure is relevant for both policy and practice. Having a better understanding of factors that may be associated with R&D project failure can aid policy design to increase the likelihood of success, and when publicly funded, it can increase the public's return on investment.

The literature to date on PIs has largely been focused on identifying their roles and responsibilities in publicly funded R&D projects. PIs have been shown to be critical to R&D projects and are generally required to perform a variety of duties in their role. In some sense, they are the gatekeepers of new technology to market, which implies they have a large responsibility for delivering social benefit. As described above, there is limited research on how characteristics of PIs are associated with R&D project failure. Given the importance of PIs to the outcome of R&D projects, it is prudent to continue to develop an understanding of how characteristics of PIs may impact R&D project failure. Further, there has not been a theoretical formulation provided in the literature to explain how characteristics of PIs are associated with project failure. Chapter 5 begins closing that gap by providing a microeconomic theoretical model to explain how characteristics of PIs may associate with R&D project failure.

First, however, Chaps. 3 and 4 provide the contextual backdrop for the empirical analysis that examines factors associated with publicly funded R&D project failure. Chapter 3 describes the history of the US Department of Energy (DOE) and its long-running history of performing R&D to innovate for achieving its mission. Chapter 4 outlines the history of the SBIR program and the DOE's involvement. The contextual background is necessary as it provides context to the program data used in the empirical analysis presented later.

Part I
Contextual Setting

Chapter 3
Publicly Funded Collaborative R&D: The Case of the US Department of Energy

Abstract The Department of Energy (DOE) is one of 15 executive departments that are the administrative arms of the executive branch of the US government, each with their own mission. The heads of each of these departments are members of the president's cabinet and are expected to advise the president in their respective areas. The DOE, in general, is focused on the environmental and energy-related needs of the United States and has a long-standing history of using technological innovation to achieve its goals. The DOE descended from one of the best-known projects that involved public–private partnerships resulting in a new technology coming to fruition. This chapter discusses the history of the DOE in an effort to provide context for the data used from that Department in the empirical analysis that follows.

3.1 Introduction

Economists generally agree that if left to markets alone, sufficient funds will not be allocated to research and development (R&D) (or funding needed to commercialize it) as it has some similar characteristics to a public good (Rosenberg, 1990). Rosenberg also explains that it is generally accepted that the social return on basic research (R&D) is generally greater than the private return, therefore R&D is typically publicly funded. In a technical report from the National Renewable Energy Laboratory within the Department of Energy (DOE), Murphy et al. (2007, p. 1), reinforce the previous notion stating, "There is a relative dearth (and gap where technologies can be stranded) of resources available to attract private sector financing for transitioning technologies out of the lab." Hence, technology and innovation policy is an important factor in shaping the R&D landscape that can impact both public and private research and development project activities and therefore economic growth. During the late 1970s and early 1980s, the United States experienced a slowdown in total factor productivity. This slowdown was the catalyst for the US Congress to implement several policies as a response to the slowdown.

M. Boyce, *Principal Investigators and R&D Failure*, Academic Entrepreneurship, Scientists, and Scientific Careers, https://doi.org/10.1007/978-3-031-43608-6_3

Five key policies enacted by Congress with the intent to foster public–private partnerships to drive innovation were:

1. The Bayh-Dole Act of 1980 (Pub. L. 96-517)
2. The Stevenson-Wydler Act of 1980 (Pub. L. 96-480)
3. The Federal Technology Transfer Act of 1986 (Pub. L. 99-502)
4. The R&D Tax Credit of 1981 (H.R. 4242)
5. The Small Business Innovation Development Act of 1982 (SBIR) (Pub. L. 97-219)

Each policy had specific measures that incentivized public–private partnerships in an effort to increase R&D and innovation in the United States. For example, the Bayh-Dole Act of 1980 was intended to provide commercial incentives for institutions conducting research even though the research may be publicly funded. The Act specified that universities, small businesses, or nonprofit institutions receiving federal funding could choose to pursue ownership of inventions instead of remaining with the government. The Stevenson-Wydler Technology Innovation Act of 1980 was intended to help the dissemination of information from the federal government to the public. While the Bayh-Dole Act was largely focused on universities and nonprofit organizations, the Steven-Wydler Act was written specifically for national laboratories. The law required federal laboratories to set aside a portion of their budget specifically for technology transfer activities. The Federal Technology Transfer Act of 1986 amended the Stevenson-Wydler Act and created the Federal Laboratory Consortium and permitted federal laboratories to enter into cooperate research and development agreements and negotiate licenses for patented technologies they invented. The R&D tax credit of 1981 was intended to incentivize firms to increase spending on research activities. Finally, The Small Business Innovation Research (SBIR) program provides public funding for small businesses to undertake research on new technologies. The SBIR was established by the US Congress through the Small Business Innovation Development Act of 1982. The Act required that qualifying governmental departments and agencies with external research programs establish and fund their own SBIR programs.

The US Department of Energy is one of the qualifying departments required to set aside a portion of its extramural research budget for its administration of the SBIR program. Hence, in this study, the SBIR program is the most relevant materialization from the policies discussed above because the data used in the empirical analysis are from the DOE's SBIR program.

The US Department of Energy (DOE) is a cabinet-level department with a stated mission "… to ensure America's security and prosperity by addressing its energy, environmental and nuclear challenges through transformative science and technology solutions" (DOE – About Us, n.d.). It was officially created through the Department of Energy Organization Act of 1977 (Public Law 95-91), which was signed into law by President Jimmy Carter on August 4, 1977.

Since its inception, the DOE has had 16 Secretaries who served under eight Presidents (see Table 3.1), yet the objectives of the agency have remained largely

Table 3.1 Department of energy secretaries

No.	Name	State of residence	Took office	Left office	Party	President(s)
1	James Schlesinger	Virginia	6-Aug-77	23-Aug-79	Republican	Jimmy Carter
2	Charles Duncan	Texas	24-Aug-79	20-Jan-81	Democratic	
3	James Edwards	South Carolina	23-Jan-81	5-Nov-82	Republican	Ronald Reagan
4	Donald Hodel	Oregon	5-Nov-82	7-Feb-85	Republican	
5	John Herrington	California	7-Feb-85	20-Jan-89	Republican	
6	James Watkins	California	1-Mar-89	20-Jan-93	Republican	George H. W. Bush
7	Hazel O'Leary	Virginia	22-Jan-93	20-Jan-97	Democratic	William Clinton
8	Federico Peña	Colorado	12-Mar-97	30-Jun-98	Democratic	
9	Bill Richardson	New Mexico	18-Aug-98	20-Jan-01	Democratic	
10	Spencer Abraham	Michigan	20-Jan-01	1-Feb-05	Republican	George W. Bush
11	Samuel Bodman	Illinois	1-Feb-05	20-Jan-09	Republican	
12	Steven Chu	California	20-Jan-09	22-Apr-13	Democratic	Barack Obama
13	Ernest Moniz	Massachusetts	21-May-13	20-Jan-17	Democratic	
14	James Richard Perry	Texas	2-Mar-17	1-Dec-19	Republican	Donald Trump
15	Dan Brouillette	Texas	1-Dec-19	20-Jan-21	Republican	
16	Jennifer Granholm	Michigan	25-Feb-21	Incumbent	Democratic	Joe Biden

Source: https://www.energy.gov/management/history/secretaries-energy

unchanged. With just over 40 years as a cabinet-level department, the DOE is still relatively new compared with other departments, though it has a rich history that can be traced back for several additional decades.

3.2 The Early History of DOE

Less than one year before Germany's invasion of Poland on September 1, 1939 (which marked the start of World War II), two German radiochemists discovered uranium.[1] Not long after the discovery, Germany stopped selling uranium and began researching its potential uses. Concerned with the possibility Germany was researching fission chain reactions using uranium with the goal of building an extremely powerful bomb, Albert Einstein drafted a letter to President Franklin D. Roosevelt

[1] This chapter draws largely from the DOE Office of History and Heritage (OHH) online resources as they provide the most complete and detailed history (https://www.osti.gov/). As such, direct quotes have been referenced and/or text have been paraphrased, but it should be emphasized that this history follows that described by the DOE OHH closely.

alerting him of his concerns. President Roosevelt received the letter on August 2, 1939, and wrote back to Einstein several days later explaining that in response to the letter, he had assembled a committee of civilian and military personnel to advise on the research of uranium. On October 21, 1939, the Advisory Committee on Uranium, headed by Lyman J. Briggs, met for the first time.

In June 1940, President Roosevelt aligned the newly constituted Committee on Uranium to be under the recently created National Defense Research Committee (NDRC). Vannevar Bush was appointed to head the NDRC, and he reorganized the Committee on Uranium into a scientific community, thus ending military membership. By eliminating the dependence on the military for funding, the NDRC was able to gain more direct access to funding for uranium research. The Committee on Uranium still held responsibility for uranium research under the NDRC and recommended that funding continue for the remainder of 1940. Bush approved the funding for the uranium research but also banned the publication of any research on uranium and barred foreign-born scientists from the committee for the sake of national security.

In 1940, German forces experienced wartime success in Europe. This success led many to believe it was only a matter of time before the United States became involved in the war. While the US Government funded uranium research in 1940, the scientific community worried about the pace of research proceeding too leisurely. Ernest O. Lawrence, director of the Radiation Laboratory at the University of California, Berkeley, was one of the most outspoken scientists concerned with the pace of uranium research. Lawrence was enthusiastic about uranium's potential and hypothesized a method to separate larger and purer amounts of uranium-235 for study. Eager to speed up this research, Lawrence contacted Karl T. Compton and Alfred L. Loomis, both of Harvard University, who were also doing work for the NDRC. Lawrence shared his views and hypotheses with Compton and Loomis. Sympathetic to Lawrence's urgency to speed up uranium research, Compton and Alfred Loomis shared Ernest Lawrence's agreement with Bush that the Committee on Uranium were moving too slowly, especially, in the face of Germany's progress.

After a meeting between Vannevar Bush and Ernest Lawrence, Bush believed Lawrence was on the right track and appointed him to be an advisor to Lyman Briggs. Bush also requested that the National Academy of Sciences, headed by Arthur Compton, review the uranium research program. In May 1941, Lawrence, in conjunction with Karl Compton, released the first report from the National Academy of Sciences (Academy) on the uranium program. The report stated that it was possible that uranium could be used for creating a radioactive weapon as early as 1943, in the event of war. However, the expected time of when a large bomb would be needed was undetermined, and it was thought that it would certainly not be needed before 1945. Bush was not appeased by the findings in the report and requested that the Academy conduct a review of the first report from an engineering standpoint. The second report, dated July 11, 1941, confirmed the results of the first report, and this was a disappointment according to Bush.

At the time Bush received the second Academy report, he had become the director for the Office of Scientific Research and Development (OSRD). The Committee on Uranium, code-named the S-1 committee, then became the OSRD Section on Uranium. While the US efforts weaponizing uranium were moving slowly, in July 1941, a draft of the so-called MAUD report, code-named for a British group of researchers, founded in 1940 to study uranium, was provided to Bush. The MAUD report was one of the most influential reports of the time as it provided details on how to build a nuclear bomb. Recognizing the importance of the MAUD report, Bush decided to strengthen the S-1 committee and requested Compton to address technical questions related to the MAUD report.

In October 1941, before Bush had received Compton's latest report, he met with President Roosevelt and Vice President Henry Wallace to discuss the positive outlook of uranium research as a result of the MAUD report. Bush received permission from the President to engage the US Army to explore construction of a bomb. The President instructed Bush to move quickly but to only conduct R&D; he also required Bush to seek additional approval to move forward with production if the situation permitted.

Just over a month before Japan bombed Pearl Harbor (December 7, 1941), Compton's committee submitted a report to Bush that confirmed the basic conclusions of the MAUD report. Bush shared this information with President Roosevelt who responded through a written note stating, "V. B. OK—returned—I think you had best keep this in your own safe FDR" (DOE – OHH, n.d.).

The United States entered World War II following the attack on Pearl Harbor. This event, coupled with the concern Germany was progressing toward the creation of an atomic weapon, sparked urgency in the federal government to support the US avenues of developing its own bomb. Through the first half of 1942, Bush reorganized the overall organizational structure of the effort to build an atomic bomb. This was in parallel with troves of new scientific information pouring in that needed to be analyzed and prioritized for the upcoming production phase. Throughout the year, progress was made toward entering the production phase of an atomic bomb. On December 9, 1942, the S-1 committee met to discuss a report prepared by Warren K. Lewis of the Massachusetts Institute of Technology (MIT), and a draft was prepared for Bush to send to President Roosevelt with recommendations for next steps. On December 28, 1942, the President approved $2 billion in spending to be used to build an atomic bomb. He also gave the Manhattan Project approval to construct the plants and facilities needed to build the bomb.

With the full support of President Roosevelt, the Manhattan Project moved forward in the efforts to build an atomic bomb. Across the country, multiple sites were working toward producing plutonium and enriched uranium. With plutonium production increasing significantly late in 1944, the facility in Los Alamos that focused on the actual design of a nuclear weapon was also making substantial progress.

On August 6, 1945, with President Harry Truman's approval, a 9700-pound uranium bomb, nicknamed "Little-Boy" was dropped on Hiroshima, Japan, with devastating effects. Just 3 days later, a plutonium bomb nicknamed "Fat-Boy" was dropped on the city of Nagasaki, again with devastating effects. On August 10, 1945, the emperor of Japan forced military leaders to offer a surrender. After some reluctance by the Japanese military leaders, Japan officially surrendered on September 2, 1945.

Following the end of World War II, the United States had to deal with the issue of how to proceed now that the world knew the forcefulness of nuclear power. Perhaps the most contentious point was whether the infrastructure developed through the Manhattan Project should be controlled by the military or by civilians. The debate was settled on August 1, 1946, when President Truman signed the Atomic Energy Act (Public Law 79-585), which went into effect on January 1, 1947. The Act established the civilian lead Atomic Energy Commission (AEC) intended to promote the "utilization of atomic energy for peaceful purposes to the maximum extent consistent with the common defense and security and with the health and safety of the public" (EPA, 2018, para. 1).

For the next few years, the AEC expanded its weapon-making sites as the so-called Cold War between the Soviet Union and the United States intensified. In August 1949, the Soviet Union detonated its first atomic weapon, which sparked President Truman to request the AEC to expedite the development of a thermonuclear weapon.

On June 25, 1950, North Korea invaded South Korea, which ignited the Korean War. The global turmoil led President Truman to approve a $1.4 billion spending bill to expand AEC facilities to produce more uranium and plutonium for nuclear weapons. On November 1, 1952, the United States detonated its first thermonuclear weapon; it created an explosion approximately 700 times greater than the uranium bomb dropped on Hiroshima.

On January 20, 1953, Dwight D. Eisenhower was inaugurated. Eight months into Eisenhower's presidency, the Soviet Union tested a bomb that was a precursor to a thermonuclear bomb. At the same time, the United States had made progress on harnessing nuclear energy for naval propulsion; the US Navy launched its first nuclear powered submarine in January 1954.

On August 30, 1954, President Eisenhower signed the Atomic Energy Act of 1954 (Public Law 83-703), which is now one of the fundamental laws of the Nuclear Regulatory Commission (NRC). The law, as it pertained to the development of civilian nuclear power programs, stated, "The development, use, and control of atomic energy shall be directed so as to promote world peace, improve the general welfare, increase the standard of living, and strengthen free competition in private enterprise" (OGC, 2013, p. 15).

Although the Soviet Union continued to make advances in its nuclear program, including the detonation of a thermonuclear weapon in 1955, a third world war was feared but never occurred. In August 1958, President Eisenhower declared a moratorium on all nuclear weapon testing effective October 31. The moratorium was in conjunction with the British and Soviets, the only other nations with nuclear programs at the time. After 3 years, the Soviet Union began extensive nuclear weapon testing again, breaking the 1958 moratorium agreement, just days later the United States resumed nuclear weapon testing. Negotiations for an international agreement among the United States, the United Kingdom, and the Soviet Union nuclear weapon programs continued for a couple of years. On August 5, 1963, the three nations signed the Limited Test Ban Treaty that prohibited underwater, atmospheric, and outer space nuclear test but did not ban underground testing.

In December 1963, plans were announced for the first nuclear power plant to be built without government aid and able to compete with conventional plants. The Jersey Central Power and Light Company was the entrepreneurial effort behind the plans. With the need for energy ever increasing the AEC focused on the development of the Liquid Metal Fast Breeder. In a report explaining the Breeder Reactors, AEC commissioner Glen Seaborg stated, "The development and use of the breeder reactor will give us an even greater amount of power-perhaps enough for thousands of years" (Mitchel & Turner, 1971, p. 1). The concern over producing enough reliable energy became a reality on November 9, 1965, when the northeastern United States experienced a major power blackout. Following the blackout in 1965, the United States and AEC continued to work toward developing a Liquid Metal Fast Breeder Reactor. By 1972, the AEC was actively working on plans with industry partners to build a Breeder Reactor.

In June 1973, President Richard Nixon recognized that nuclear research and development activities had propagated since the Manhattan Project. The energy crisis of the 1970s was also underway and beginning to hit US consumers. As such, the President requested Dixy Lee Ray, Chairman of the AEC at the time, to conduct a review of all energy-related research and development activities and to recommend a unified national program. Later that month, President Nixon established the Energy Policy Office. Within 6 months, President Nixon replaced the Energy Policy Office with the Federal Energy Office (FEO). The new office was given the responsibility of controlling the price of oil and gasoline and rationing petroleum supplies to refiners. Only a few months later, the President replaced the FEO with the Federal Energy Administration through the Federal Administration Act of 1974 (Public Law 93-275).

Shortly after President Gerald R. Ford took office, with the energy crisis in full swing, he signed the Energy Reorganization Act of 1974 (Public 93-438), which replaced the AEC with the NRC and the Energy Research and Development Administration (ERDA). The NRC was tasked with regulating civilian use of nuclear materials and the ERDA administered research and development programs related to the use of various energy sources. Also, during President Ford's administration, construction began on the Trans-Alaska Pipeline; this was the largest private construction project in American history at the time. Along with the construction of the pipeline came debate over the potential negative environmental impacts of the 800-mile pipeline. These concerns plus the desire to create an oil reserve prompted the enactment of the Energy Policy and Conservation Act (Public Law 94-163), signed into law by President Ford on December 22, 1975.

Four months into President Carter's presidency, in April 1977, he delivered his first major speech on energy. Through his speech, he unveiled a plan to establish an energy department. In President Carter's address to the nation he stated, "The energy crisis has not yet overwhelmed us, but it will if we do not act quickly. It's a problem that we will not be able to solve in the next few years, and it's likely to get progressively worse through the rest of this century" (Carter, 1977). Following the

President's address, in August 1977, President Carter signed the Department of Energy Organization Act (Public Law 95-91), thus abolishing the Federal Energy Administration and the ERDA. The Department of Energy was officially formed on October 1, 1977, consolidating many entities from several departments and agencies. In accordance with President Carter's plan, the Act was intended, as stated in Public Law 95-91, "To establish a Department of Energy in the executive branch by the reorganization of energy functions within the Federal Government in order to secure effective management to assure a coordinated national energy policy, and for other purposes." The DOE was also given responsibility over the US nuclear weapons program, whose lineage can be traced back directly to the Manhattan Project.

3.3 DOE Is Formalized

James R. Schlesinger was appointed to be the first Secretary of Energy. Since the DOE brought together several, largely independent agencies and offices, Schlesinger's first task was to meld all of them into a unified department. The new department consisted of approximately 20,000 employees with an annual budget of $10.4 billion. The newly created department, "despite its diverse origins, was structured to allow for the continuity of programs and functions from predecessor organizations while blending their expertise into new management teams" (Fehner & Holl, 1994, p. 23). The Department also inherited, mostly from the AEC through the ERDA, many regional and field offices, laboratories, research centers, and university programs from predecessor agencies. Thus, the Department of Energy was not intended to reduce resources through consolidation but was formed to build on previous efforts with a more strategic structure.

Through the end of 1977, President Carter focused on creating legislation that would bring his administration's energy policies together to form a National Energy Plan. However, public opinion was not favorable to President Carter's plan, and special interest groups opposed to the plan were successful in preventing any legislation to be enacted. Not creating an official National Energy plan was a big disappointment for the Carter Administration, so they continued to push a policy forward in the first half of 1978. On November 9, 1978, President Carter was finally successful in creating a National Energy Plan when he signed into law the National Energy Act of 1978. The legislation was substantial; it consisted of five major Public Laws:

1. The National Energy Conservation Policy Act (Public Law 95-619).
2. The Powerplant and Industrial Fuel Use Act (Public Law 95-620).
3. The Public Utilities Regulatory Policy Act (Public Law 95-617).
4. The Energy Tax Act (Public Law 95-618).
5. The Natural Gas Policy Act (Public Law 95-621).

The National Energy Act was considered a success for President Carter and the DOE. With an official National Energy Plan in place, Secretary Schlesinger and the DOE had a newfound charter and the funding to work toward their new goals.

Successfully passing the National Energy Act was only a small step toward effective energy policies. The DOE submitted its first comprehensive budget request for the year 1980, as opposed to an aggregation of legacy agency budget requests. With an $8.4 billion budget and thousands of employees, the DOE was not a simple organization to manage. Well over a year after the agency was established, the organization was still trying to find its rhythm, which thus created many critics of the DOE.

As the DOE was trying to get its footing and iron out its initial start-up issues, the ongoing energy crisis took an unfortunate turn for the worse. In 1979, the world began to experience an oil shortage driven by turmoil in Iran leading to the cessation of oil exports from the country. As energy prices spiked, Secretary Schlesinger realized there would be no easy solution to the crisis. Considerations were given to voluntary conservation measures; however, the situation continued to deteriorate. As if the energy crisis was not straining the relatively new DOE enough, another challenge arose with an accident at a nuclear power plant in Harrisburg, Pennsylvania. Failures in the system resulted in the release of radioactive material that cost about $1 billion to cleanup (New York Times, 1993). Though the world was facing oil shortages, President Carter made it clear that the use of nuclear power should be a last resort.

The Energy Crisis intensified during the first half of 1979, and in some cases, violence broke out due to the need for rationed gasoline. President Carter addressed the nation explaining that not one single factor caused the crisis and that the American people must speak out to combat the oil companies' special interests in high prices. The DOE created teams of auditors to check up on refiners and individual service stations to try and enforce gasoline ceiling price regulations. Congress was not behind the President's methods in resolving the energy the crisis; however, President Carter persisted that renewable energy would help alleviate the dependence on foreign oil. Nonetheless, a majority of Americans at the time believed the energy crisis was artificially contrived by the government, oil companies, and oil-producing nations. The skepticism did not bode well for the DOE's reputation.

Secretary Schlesinger resigned his office on July 16, 1979. President Carter then selected Charles W Duncan, Jr. to be the second Secretary of Energy. Secretary Duncan believed that the business of energy belonged in the private sector and that the proper role of government was the effective allocation of public resources. He also suggested that the government should provide proper incentives to private enterprise to help transition to an energy-diversified economy from one dependent on oil. The DOE was still in its infancy when Secretary Duncan took the reins; thus, part of his task was to improve the management structure of the department. The original structure of the department was based on the evolution of technologies from research and development through commercialization. Secretary Duncan reorganized the department according to respective fuel or technology types, a more traditional structure.

During the summer of 1979, the energy crisis began to subside. Americans were consuming less oil driven by an economic downturn and high gas prices. The following year, a presidential election took place. President Carter and Republican presidential candidate Ronald Reagan largely avoided making the nation's energy issues a major issue in their campaigns. Reagan, the victor, criticized the past administration's energy policy and advocated for the abolishment of the DOE. Following Reagan's election, he nominated James B. Edwards as the third Secretary of Energy. Edwards and the Reagan Administration quickly refocused the DOE's mission. As opposed to President Carter's policies, they did not want the government involved in activities that the private sector and markets could manage, such as price controls and regulations that slowed domestic production. On February 25, 1981, Secretary Edwards announced an organizational structure change that would permit the DOE to focus more on research, development, and production.

Only 4 years after the DOE was established, the agency had become to many a symbol of the ineffectiveness of government overreach. Secretary Edwards was in favor of dismantling the DOE and reorganizing a more research-focused administration under the Department of Commerce (DoC). When the Department of Energy Organization Act was created, it included a clause that required the President to submit a review of the Department to Congress before the Department could be dismantled. Unfortunately for President Reagan, Congress gave the DOE sound marks in achieving its goals and would not permit the complete abolishment of the department. By 1982, the nation's energy situation had improved, and Secretary Edward touted that free markets and little government intervention were behind the change. Secretary Edwards had basically declared the end of the energy crisis and shortly thereafter gave his resignation to President Reagan.

On October 5, 1982, Donald P. Hodel was named the fourth Secretary of Energy. Secretary Hodel held a similar sentiment as his immediate predecessor and tried to reorganize the DOE under the Department of Commerce (DoC). One main sticking point with his plan was that Congress did not believe the nuclear weapons program, maintained by the DOE, should be a program under the DoC. With the DOE remaining intact, Secretary Hodel largely carried out President Reagan's objectives deregulating energy markets and reducing the department's personnel. During Secretary Hodel's tenure as secretary, one of the most hotly debated issues revolved around the government's role in energy research and development. Secretary Hodel believed that research and development should lie in the private sector but that government should intervene when the research was too expensive for the private sector to undertake but had the potential for large benefits.

Ten days before President Reagan was inaugurated for his second term, he appointed John S. Herrington as the fifth secretary of energy. Secretary Herrington's priorities were roughly in line with those of his predecessor. Secretary Herrington suggested energy policy should consist of three objectives: energy stability, energy security, and energy strength. Energy stability and energy security had been the focus of the DOE since the energy crisis in 1973, so Secretary Herrington wanted to focus on energy strength to make progress toward his view of an effective energy policy. By the mid-1980s, the United States had become much more efficient in how it consumed energy. Energy conservation was no longer just a slogan, it had become an energy resource. Conservation, nuclear energy, and coal energy were the three components Secretary Herrington believed would help the United States achieve energy strength.

Secretary Herrington was also a strong proponent of government-funded basic research. In 1986 and 1987, his views were appeased when major breakthroughs in electric technology efficiency provided the means to achieve superconductivity (zero electrical resistance) at lower costs. President Reagan had become impressed with the potential for basic scientific research to provide life-changing innovations. Secretary Herrington believed this new breakthrough showcased how the president's energy policy, and the DOE worked at their best. The key formula was the exchange of information and ideas among universities, private industry, scientific laboratories, and government. In fact, in a press conference, President Reagan suggested the Superconductivity initiative demonstrated the administration's policy for "the swift transfer of technology and technical information from the government to the private sector" (White House, 1987, p. 1).

As President Reagan's presidency concluded, the DOE was much different from when it was created. President Reagan was not successful at eliminating the department, but Secretary Herrington believed the current state was much more appealing to the president. The DOE was no longer focused on regulatory functions, instead was supporting R&D and weapons facilities. After a decade since the DOE's inception, arguments over energy policy have largely subsided.

Over the next decade, two more presidents would hold the office; George H. W. Bush from January 20, 1989, to January 20, 1993, and William J. Clinton from January 20, 1993, to January 20, 2001. Under President Bush, Admiral James Watkins was named the Secretary of Energy. Early in his tenure as secretary of energy, Secretary Watkins's biggest challenge was handling environmental problems caused by nuclear waste from the nuclear weapons complexes. This problem was essentially solved through the enactment of the Waste Isolation Pilot Plant Land Withdrawal Act (Public Law 102-579). The Act was signed into law in 1992 and designated a single location to store defense-related waste. Apart from the issues of nuclear waste, the DOE under the Bush administration largely echoed the energy policy of the Reagan administration. The Bush administration stressed that a diversified energy portfolio, consisting of coal, nuclear power, oil and natural gas, renewables, alternative fuels, and conservation was needed for the nation's energy security. Along these lines, President Bush presented the National Energy Strategy to the nation, including Congress, which was touted as a pro-production strategy. Secretary Watkins suggested the strategy was a first of its kind to provide energy security and environmental quality through deregulation, free markets, and investment in research and development. Although the strategy had mixed opinions, it did not cause any major political waves and was largely implemented through the passage of the Energy Policy Act of 1992 (Public Law 102-486).

When President Clinton took office, the Cold War had ended, thus the DOE's defense-related activities were giving way to environmental restoration and waste management activities. Energy itself had largely fallen off the radar of most Americans, but how energy interacted with the environment had become a major concern. Early in his presidency, President Clinton made it clear that the secretaries of energy and commerce were crucial to his economic policy and that the DOE would be central to the administration's policy goals. According to the Clinton Administration and echoed by Hazel Rollins O'Leary, President Clinton's first appointed Secretary of Energy, the DOE would be focused on energy policy that relied more heavily on American natural gas, conservation through efficiency gains, alternative fuels, and more consideration for the environment. Over the course of President Clinton's first term as President, he and the DOE continued to push their environmental policy forward. By the start of President Clinton's second term, a debate on whether the DOE should focus on applied or basic research was underway. Secretary O'Leary pushed the department toward applied research and thought technology transfer from public laboratories to the private sector would be key to maintaining the public laboratories following the end of the Cold War. Although the Bush Administration also pushed for technology transfer, under Secretary O'Leary, and in just her first year, the number of Cooperative Research and Development Agreements (CRADAs) negotiated between the DOE and academia, industry, and others doubled those negotiated while under Secretary Watkins' tenure.

Since the Cold War had ended, and given the focus on the environment, nuclear power had lost major attention by the DOE except for the ongoing nuclear waste cleanup. This changed with the election of President George W. Bush in 2001. President Bush was a proponent of both nuclear power and the oil industry, so during his administration, the DOE had a renewed focus on nuclear and oil energy. The capstone of Bush's energy policy was the Energy Policy Act of 2005 (Public Law 109-58) that provided a broad range of subsidies for nuclear and oil companies.

When President Barack Obama was elected in 2008, he brought a renewed attention to how energy impacts the environment. "*Clean energy*" was the term coined to represent energy sources that have a much smaller negative impact to the environment than coal, oil, and nuclear. Under Obama Administration's energy policy, several investments were made in clean energy including tax credits for the solar and wind industries, funding for a smart grid, and subsidies to make low-income homes more energy efficient. The funding for these investments were provided through the American Recovery and Reinvestment Act of 2009 (Public Law 111-5). As stated in the Act, the purpose of this policy was for "Making supplemental appropriations for job preservation and creation, infrastructure investment, energy efficiency and science, assistance to the unemployed, and State and local fiscal stabilization, for the fiscal year ending September 30, 2009, and for other purposes."

3.4 Current Organization of the DOE and Support for PIs

As of 2022, the DOE is led by Secretary Jennifer Granholm, the 16th Secretary of Energy. The Department is operating relatively similar to past year's departments under prior administrations. The Department's 2021 fiscal year budget appropriation of $39.6 billion consists of six major programs. The National Nuclear Security Administration is the largest program by funded amount of $19.7 billion of which

$12.5 billion is designated for weapon activities. The next largest funding amount of $7.3 billion is from the Environmental Management program to be used for continuing the cleanup of waste generated from 50 years of nuclear weapon development and public nuclear energy research. With an appropriation of slightly more than $7.0 billion, the Office of Science uses funding to continue the DOE's long history of early-stage research and development in an effort to stay at the forefront of scientific innovation. The fourth major category of the budget comes from the Energy program. The Energy program uses its funding to promote technologies that will make the American energy supply more reliable, affordable, and efficient. The department's budget by major organizations over the past several years is presented in Fig. 3.1. Additionally, Fig. 3.2 presents the extramural research budget of the DOE over the past several years.

Consistent with its origins, the DOE continues to engage in innovative research. According to its most recently released strategic plan (DOE, 2014, p. 10), "DOE will continue to pursue scientific discoveries that lay the technological foundation to extend our understanding of nature and create new technologies that support DOE's energy, environment, and security missions." In an effort to achieve the DOE's plan to perpetuate innovation, the Department provides support for principal investigators (PIs) and other scientists to further their research.

For example, the DOE's Office of Science funds a wide range of research projects through its various programs, including the Biological and Environmental Research, Basic Energy Sciences, Advanced Scientific Computing Research, and Fusion Energy Sciences programs. PIs can submit proposals for funding in areas that align with the goals of these programs. Also, the Office of Energy Efficiency and Renewable Energy (EERE) sponsors the Technology-to-Market Program that supports PIs through funding early-stage R&D in areas such as solar energy, bioenergy, and energy efficient building with the goal of commercialization. And as

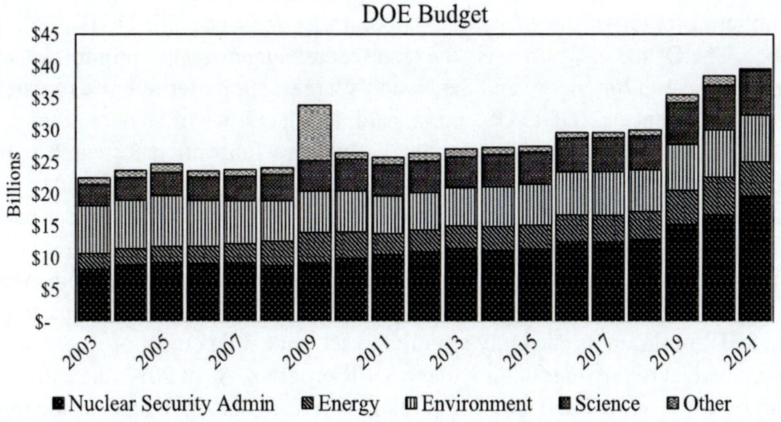

Fig. 3.1 Department of energy budget by major organizations. (Source: https://www.energy.gov/cfo/listings/budget-justification-supporting-documents)

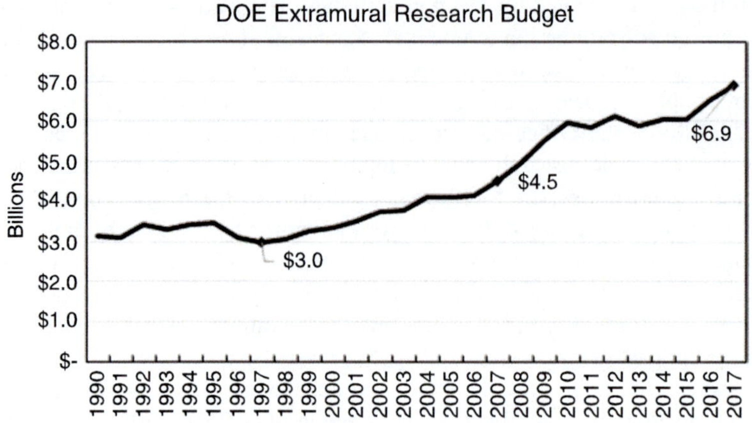

Fig. 3.2 DOE Extramural research budget

mentioned previously, the DOE participates in the SBIR program that provides awards for small businesses and their PIs.

More generally, 17 DOE national laboratories are used to advance the Department's mission and aid in translating basic research to innovation. The national laboratories are critical to scientific innovation and possess unique instruments and facilities not found anywhere else in the world. Access to the national laboratories is another way the DOE supports scientists. The DOE, through its Office of Science, also funds user facilities that provide researchers, including PIs, from academia, industry, and the government with advanced scientific tools to perform new scientific research. In fiscal year 2015, over 32,000 researchers spanning all 50 states and Washington DC utilized a user facility (DOE, 2015).

Furthermore, ten of the 17 national laboratories fall under the DOE's Office of Science. The Office of Science is "the lead federal agency supporting fundamental scientific research for energy and the Nation's largest supporter of basic research in the physical sciences," (DOE-OS, n.d.-a, para. 1). The Office of Science administers eight programs, which includes the Small Business Innovation Research (SBIR) program; the DOE is currently one of 11 federal agencies that participates in the program. In addition to the Office of Science, seven additional DOE offices participate in the SBIR program.

Table 3.2 shows the DOE offices that participate and the links to their mission statements. The DOE, consistent with the other ten federal agencies that participate in the SBIR program, is currently required to set aside 3.2 percent of its extramural research budget to provide funding to the SBIR program. As of 2017, the extramural research budget at the DOE was $6.9 billion, therefore, the amount set aside for the SBIR program was $223.7 million.

Table 3.2 Department of Energy SBIR program participating offices

DOE office	Link to mission statement
Office of Cybersecurity, energy security, and emergency response	https://www.energy.gov/ceser/office-cybersecurity-energy-security-and-emergency-response
Office of Electricity	https://www.energy.gov/oe/office-electricity
Office of Energy Efficiency and Renewable Energy	https://www.energy.gov/eere/office-energy-efficiency-renewable-energy
Office of Environmental Management	https://www.energy.gov/em/office-environmental-management\
Office of Fossil Energy	https://www.energy.gov/fe/office-fossil-energy\
Office of Defense Nuclear Nonproliferation R&D	https://www.energy.gov/nnsa/missions/nonproliferation\
Office of Nuclear Energy	https://www.energy.gov/ne/office-nuclear-energy\
Office of Science	https://science.energy.gov/
Office of Advanced Scientific Computing Research	https://science.energy.gov/ascr/
Office of Basic Energy Sciences	https://science.energy.gov/bes/
Office of Biological and Environmental Research	https://science.energy.gov/ber/
Office of Fusion Energy Science	https://science.energy.gov/fes/
Office of High Energy Physics	https://science.energy.gov/hep/
Office of Nuclear Physics	https://science.energy.gov/np/

3.5 Conclusion

The DOE has a long history of engaging in public–private partnerships to support its needs for technological advancement and innovation to achieve its mission. Currently, the DOE supports R&D through access to state-of-the-art equipment at national laboratories and user facilities and funding for scientists through its various programs. However, in the early 1980s, there was a concern that the United States was falling behind other nations in its ability to innovate. Therefore, in an effort to stimulate innovation, policy makers created several policies that incentivized R&D, including the SBIR. The DOE's SBIR program provides awards to small businesses that are performing R&D with the intent to commercialize their technology. The data in the empirical analysis provided later in this monograph comes from the DOE's SBIR program. These data include firm and project characteristics as well as demographic information on the firm's PI. Hence, given the relevance of the SBIR program to this piece, Chap. 4 provides a more thorough discussion of the SBIR program.

Chapter 4
Legislative History of the SBIR Program

Abstract This chapter provides an overview of the legislative history of the SBIR program along with the Department of Energy (DOE)'s historical involvement in the program. It also provides a review of the relevant economics literature related to the Small Business Innovation Research (SBIR) program. One conclusion from the literature review is that there is a body of research that identifies covariates with the programmatic success of an SBIR project when measured in terms of the likelihood of project commercialization. Absent from that literature is a detailed theoretical analysis of SBIR project failure, project failure being the focus of this monograph.

4.1 Introduction

The Small Business Innovation Research (SBIR) program is an important program in the United States that supports innovative research in small firms through a three-phase award program. According to SBIR-STTR (Small Business Technology Transfer) (2020, p. 1), "These two initiatives, the SBIR and STTR programs are sometimes referred to as the nation's largest source of early stage/high-risk funding for start-ups and small business." As such, the SBIR program is central to the empirical analysis provided in this monograph because it provides a real-world example of small entrepreneurial firms engaging in research and development (R&D). Each firm that applies for an SBIR award is required to have a principal investigator, therefore data from the SBIR program allow the theoretical model of the probability of failure, proposed in a subsequent chapter, to be tested empirically.

The majority of SBIR program awards are administered through government agencies such as the Department of Energy (DOE) that have sizable extramural research budgets. These agencies use funds, that are set aside specifically for their administration of the SBIR program, to provide funding to small firms conducting R&D in a field that aligns with the agency's mission. The empirical analysis that is provided in this monograph uses data from the DOE's SBIR program, specifically data on Phase II award recipients.

M. Boyce, *Principal Investigators and R&D Failure*, Academic Entrepreneurship, Scientists, and Scientific Careers, https://doi.org/10.1007/978-3-031-43608-6_4

The remainder of this chapter is organized as follows: First, a discussion of the legislative history of the SBIR program is provided, followed by an analysis of the empirical literature on the SBIR program. Finally, a brief discussion of the DOE's participation in the SBIR program is provided.

4.2 The SBIR Program

Earlier than the legislative birth of the SBIR program, Congress recognized that small businesses were important to economic growth, and that they may require special treatment to remain competitive with larger enterprises. In 1953, Congress established the Small Business Administration (SBA) through the Small Business Act of 1953 (Public Law 85-536). As stated in the legislation the intent of the Act of 1953 is to:

> … aid, counsel, assist, and protect insofar as is possible the interests of small-business concerns in order to preserve free competitive enterprise, to insure that a fair proportion of the total purchases and contracts for supplies and services for the Government be placed with small-business enterprises, and to maintain and strengthen the overall economy of the Nation.

Over the next several years following the enactment of the Act of 1953, Congress provided assistance to small businesses through four primary programmatic functions: access to capital (e.g., direct business loans and guarantees on bank loans), education and counseling on the entrepreneurial process, government contracting (e.g., helping small business get government procurement contracts), and advocacy.

By the late 1970s, Congress was concerned that the United States was becoming less competitive in the global economy. There was also growing evidence that small businesses were becoming more important in job creation and the innovation process (NRC, 2008). For example, in 1979, a report from Massachusetts Institute of Technology (MIT's) Neighborhood and Regional Change program was published that highlighted the significance of small business in the job creation process (Birch, 1979). Birch found that (1979, p. 29):

> On average about 60 percent of all jobs in the U.S. are generated by firms with 20 or fewer employees, about 50 percent of all jobs are created by independent small entrepreneurs. Large firms (those with over 500 employees) generate less than 15 percent of all net new jobs.

Small business's majority share in job creation at the time was, at least partially, driven by their entrepreneurial ability to adjust to the changing global economy. The early 1980s have been referred to as the Entrepreneurial Economy and according to Link and Scott (2013, p. 14), "… smaller firms have a greater ability to be innovative, or to adopt and adapt others' new technologies and ideas, and thus quickly and efficiently appropriate investments in new knowledge that are made externally."

Although Congress had long recognized small businesses as being important for stimulating economic growth plus the then growing evidence of their importance in job creation, there was growing concern that this sector was being neglected. In 1977, an SBIR prototype program at the National Science Foundation (NSF), designed by Roland Tibbetts, began. Tibbetts, who worked in the private sector for several years before joining the NSF, foresaw a three-phase program structure:

> … in order to foster the R&D of small, high-tech businesses and push them to realize their commercial potential. He believed these firms were instrumental in converting government R&D into public benefit through technological innovation and commercial applications, therefore stimulating aggregate economic growth. (SBIR-STTR, 2018b, para. 3)

Following the success of the prototype program, in 1982, Congress amended the 1953 Act, with the intent: "… to strengthen the role of the small, innovative firms in federally funded research and development …" In the 1982 amendment to the Act of 1953, Congress stated:

1. technological innovation creates jobs, increases productivity, competition, and economic growth, and is a valuable counterforce to inflation and the United States balance-of-payments deficit;
2. while small business is the principal source of significant innovations in the Nation, the vast majority of federally funded research and development is conducted by large businesses, universities, and Government laboratories; and,
3. small businesses are among the most cost-effective performers of research and development and are particularly capable of developing research and development results into new products.

The amendment to the Small Business Administration Act, specifically the Small Business Innovation Development Act of 1982 (Public Law 97-219), established the SBIR program. The objectives of the program are:

1. To stimulate technological innovation.
2. To use small business to meet federal research and development needs.
3. To foster and encourage participation by minority and disadvantaged persons in technological innovation.
4. To increase private sector commercialization of innovations derived from federal research and development.

To be eligible for funding from the SBIR program, a firm must meet the following criteria:

1. Employ fewer than 500 employees.
2. Be independently owned with at least 51 percent ownership by US citizens or lawfully admitted permanent resident aliens.
3. Not be the dominant firm in the proposed projects field.
4. Be a for-profit organization.
5. Be the primary employment of the project's principal investigator.

The proposing firm must also perform at least two third of the R&D work in Phase I and at least one-half in Phase II.

The Amendment to the 1982 Act required that each federal agency with external research program budgets greater than $100 million for fiscal year 1982, and any subsequent year, establish and fund their own SBIR programs. Regarding the amount each agency must reserve for the SBIR program, the Act states each qualifying agency:

> … shall expend not less than 0.2 per centum of its extramural budget in fiscal year 1983 or in such subsequent fiscal year as the agency has such budget, not less than 0.6 per centum of such budget in the second fiscal year thereafter; not less than 1 per centum of such budget in the third fiscal year thereafter, and not less than 1.25 per centum of such budget in all subsequent fiscal years with small business concerns specifically in connection with a small business innovation research program which meets the requirements of the Small Business Innovation Development Act of 1982 and regulations issued thereunder.

To achieve the program objectives, a two-phase competitive research awards program was established. The Phase I awards are a relatively small amount and generally cover a six-month period. The intent of the Phase I award is to aid domestic businesses in analyzing the technical merit (i.e., proof of concept) and feasibility of commercializing the proposed R&D effort. The Phase II award, which is based on the success of the Phase I research, is larger, and the research typically lasts for about two years. The intent of the second phase is to continue the R&D efforts initiated in the first phase, ideally resulting in commercializable output. Only Phase I SBIR projects are eligible to apply for Phase II awards, but not all Phase I-awarded firms receive a Phase II award. Beyond the two phases officially supported by the SBIR program, a third phase is defined, Phase III, which is unfunded by the SBIR program; it is intended to define that period of time when the small business pursues commercialization of the efforts from the first two phases. The expectation is that the funded firm will seek third-party financial support for this final phase.

The 1982 Act was not permanent, therefore, it has been subject to several reauthorizations and a few amendments that modified the structure of the SBIR program. The first reauthorization was in 1986 through the Department of Defense Appropriation Act of 1986 (Public Law 99-443), which extended the 1982 Act through 1992. In 1992, the Small Business Research and Development Enactment Act (Public Law 102-564) reauthorized the SBIR program until 2000. The 1992 reauthorization increased the maximum set-aside rate from 1.25 percent to 2.50 percent, reemphasized the goal of increasing private sector commercialization of SBIR-funded technologies, increase Phase I awards to $100,000 and Phase II awards to $750,000, and broadened the third program objective to include women. The Small Business Reauthorization Act of 2000 (Public Law 106-554) reauthorized the SBIR program until September 30, 2008, without modifications to the required set-aside rates or award amounts.

In 2008, Congress failed to reauthorize the SBIR program by the September 30 deadline date; however, Congress did extend it through March 20, 2009, through Public Law 110-235. On March 19, 2009, Public Law 110-10 extended the program through July 31, 2009, and it was again extended until September 30, 2009, by a Senate continuing resolution (S.1513). House bill (H.R. 3614) was passed on

September 23, 2009, which extended the SBIR program until October 31, 2009. On March 30, 2010, the Small Business Administration amended the SBIR Policy Directive. This amendment increased the available Phase I award amount to $150,000 and Phase II award amount to $1,000,000 as proposed in the failed Senate bill, S. 3029, from September 2008.

Senate bill, S. 1929, extended the program through April 30, 2010, and the Senate continued the trend with the short-term extension via S. 3253, which extended the program until July 31, 2010. The program was then extended again through September 30, 2010, by the House (H.R. 5849) and then again through January 31, 2011 by the Senate (S. 3839). The House followed suit by extending the SBIR program until May 31, 2011 (H.R. 366). On May 31, 2011, the Senate temporarily extended the program again through September 30, 2011 (S. 1802). Two consecutive house bills extended the program, first to November 18 (H.R. 2608) then to December 16, 2011 (H.R. 2112). H.R. 2112 was the last of the temporary extensions resulting from Congress's failure to reauthorize the SBIR program in 2008.

On December 31, 2011, President Obama signed the National Defense Authorization Act of 2012 (Public Law 112-81), which codified the reauthorization of the SBIR program through September 30, 2017. On December 23, 2016, President Obama signed the National Defense Authorization Act of 2017, which reauthorized the SBIR program through September 22, 2022. Table 4.1 summarizes the key legislation that has kept the SBIR program running since its inception.

Currently, there are 11 agencies participating in the SBIR program, and the set-aside rate is now 3.2 percent through 2022. The agencies participating in the program are the Department of Agriculture, Department of Commerce, Department of Defense, Department of Education, Department of Energy, Department of Health and Human Services, Department of Homeland Security, Department of Transportation, Environmental Protection Agency, National Aeronautics and Space Administration, and National Science Foundation. Of the 11 participating agencies, five account for approximately 97 percent of all SBIR funding, with the Department of Defense committing the largest share.

Table 4.1 SBIR program legislation

Legislation	Public law	Authorization period
Small Business Innovation Development Act of 1982	97-219	1982–1986
Department of Defense Appropriation Act of 1986	99-443	1986–1992
Small Business Research and Development Enactment Act of 1992	102-564	1992–2000
Small Business Reauthorization Act of 2000	106-554	2000–2008
Short-term extensions	–	2008–2012
National Defense Authorization Act of 2012	112-81	2012–2017
National Defense Authorization Act of 2017	114-328	2017–2022

Source: Based on Link and Scott (2013) Table 5.1 with additional information

4.3 The Empirical Literature on the SBIR Program

As part of the SBIR Reauthorization Act of 2000 (S. 493, Sec 307), the National Research Council (NRC)[1] received a mandate to report on assessments of the SBIR program, to each funding agency, and required updates every four years. The intent of the assessments is to understand the economic benefits of the program. Prior to the 2000 mandate, there was a relatively limited body of literature that assessed the program (National Academies, 2016), especially given the size of the program. Since the first round of mandated assessments, the literature has grown substantially. Today, there is a significant literature that has studied many facets of the SBIR program. Areas that have been studied include general policy evaluations (e.g., Link & Scott, 2009; Gicheva & Link, 2013), the SBIR programs association with employment growth (e.g., Link &Scott, 2012a, b), although creating employment opportunities is not part of the programs stated mission, spillover benefits (e.g., Audretsch et al., 2002; Allen et al., 2012), and the mechanism of the award system itself (e.g., Bhattacharya, 2018). In a recent assessment of the SBIR program at the DOE, the National Academies (2020) expanded the focus of its assessment relative to the 2016 review (Academy, 2016). The recent review stated the objectives of the review were to examine (Fig. 4.1):

1. A range of economic impacts including, to the extent practicable, the number of jobs created by these programs.
2. The role of SBIR/STTR programs in stimulating technological innovation and contributing to DOE's research and development needs.
3. Collaborations created between small businesses and research institutions on account of the programs.

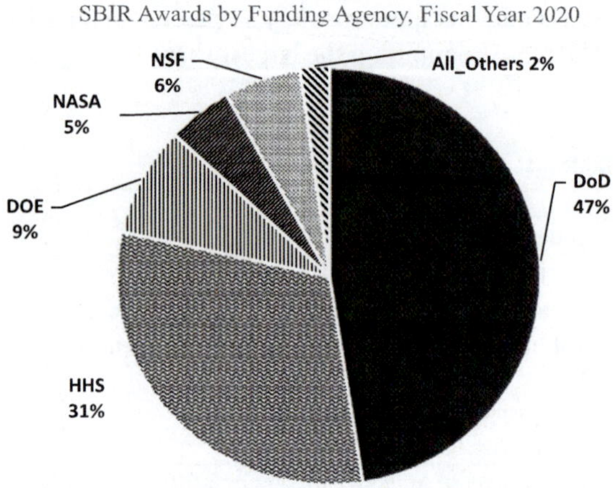

Fig. 4.1 SBIR program funding by major department. (Source: https://www.sbir.gov/node/735727?view_by=Year)

[1] NRC refers to the National Research Council from this point forward as opposed to the National Regulatory Commission, unless otherwise noted.

4. The effectiveness of DOE's SBIR/STTR award-selection process and commercialization assistance.
5. Ways to improve outreach efforts to SBIR/STTR applicants, particularly to increase applications from small businesses that are (1) new to the programs, (2) from underrepresented states, (3) woman-owned, or (4) minority-owned. (Academy, 2020, p. 22).

As it relates to this book, two areas of the empirical literature are focused on, namely the commercialization success or failure of funded projects and spillover benefits from the SBIR program.

Government intervention in markets should, theoretically, be designed to overcome or ameliorate a market failure. The market failure in the case of small firm technology development and commercialization is that there is an underinvestment in private R&D. Invoking market failure as the justification for the SBIR program suggests that the net social benefits associated with the program are positive. If public funding for private R&D increases private benefits to be at least equal to private costs, then society will be able to realize the social benefit, otherwise, the R&D effort should not be undertaken.

Concerned with estimating the social rate of return from the SBIR program, Audretsch et al. (2002) analyzed data from interviews conducted by Link and Scott (2000) of Department of Defense (DoD) SBIR award recipients for 44 projects in 43 companies. Audretsch et al. (2002) found that, on average, across the 44 projects for which they had data, the expected private rate of return without SBIR funding was 25 percent, while the lower bound of the expected social rate of return was 84 percent. The gap between the private return and social return is driven by a lower social required rate of return (hurdle rate) and difficulties appropriating the return on investment of the R&D. The private hurdle rate is higher than the social hurdle rate, therefore, the social return will always be greater than the private return for R&D projects. Allen et al. (2012) also performed an analysis to estimate the social benefit from the SBIR program. Using data from the top five government agency SBIR program participants, Allen et al. (2012) estimated the producer and consumer surplus from the population of Phase II–funded awards between 1992 and 2001. They assumed the firm bringing a new technology to market would enjoy monopoly power for some time until others entered the market. Using these surplus estimates, the authors found the demand elasticities associated with a cost-benefit ratio of unity for the projects. All five agencies had elastic demands at a cost-benefit ratio equal to 1, and the DOE's elasticity was estimated to be 1.518. Allen et al. (2012) state that there is no known elasticity of demand; however, it is likely the actual demand elasticity is lower than 1.518 due to the firm enjoying monopolistic power for some time. This implies a social benefit-to-cost ratio greater than 1, hence a positive spillover from the SBIR program.

While Audretsch et al. (2002) and Allen et al. (2012) were concerned with identifying the presence and measure of social benefit from the SBIR program, Link and Ruhm (2009a) discuss the source of the societal spillover by analyzing the vocational background of entrepreneurs who received SBIR funding from the National

Institute of Health (NIH). Link and Ruhm (2009a) explained that human capital from R&D projects can add to society's benefit through two competing channels: one public (research publications) and one private (patents). They found that entrepreneurs with business backgrounds have a 9.5 percentage point decrease in the probability that the intellectual capital will be disseminated through publications only. They also found that when an entrepreneur partners with a university, there is an 11.3 percentage point increase in the probability that publication is the only source to propagate intellectual capital.

In a study using data from the NIH SBIR program, Toole and Czarnitzki (2009) analyze the association between the human capital of scientists with a biomedical academic background and firm performance. The authors found, "that academic scientific knowledge has an institutional specificity that limits its transferability or applicability to a commercialization environment …" (Toole and Czarnitzki, 2009, p. 112). This result is consistent with Link and Ruhm (2009a) and further points to the idea that the source of the knowledge spillover is more likely to be from publications than from patents. Toole and Czarnitzki (2009) further explain that the former result is most prevalent when the principal investigator is more adept at basic research or idea generation as opposed to applied research.

Commercialization is only one of the objectives of the SBIR program; however, discussions about the success of the program often cite commercialization activity as the success metric. Phase III, albeit a phase without official support from the SBIR program, is when commercialization is supposed to occur. The program does not award funds past Phase II so any additional capital needed to bring the R&D to market must come from alternative sources than the SBIR program. Therefore, it is quite possible that a small business will be awarded both Phase I and II awards but fail to bring their efforts to market.

In a study using data from NIH SBIR Phase II–funded projects, Link and Ruhm (2009b) estimated the probability a project will commercialize as a function of receiving additional funding plus other control variables. The dependent variable was a binary indicator equal to one if the firm commercialized its research. The authors found that if a firm received additional funding above the Phase I and II awards, then the probability of commercializing was approximately 35 percentage points higher. Link and Ruhm (2009b) also found that when a university is involved with the project, the probability of commercializing increases by 12 percentage points over those that did not have university involvement.

In a similar study, Siegel and Wessner (2012) were concerned with how university involvement is associated with the commercialization success of SBIR award recipients. Instead of a binary indicator of success as mentioned previously, Siegel and Wessner (2012) define seven measures related to aspects of commercialization: (1) actual sales, (2) expected sales, (3) new employees, (4) patents applied for, (5) copyrights applied for, (6) trademarks applied for, and (7) licensing agreements consummated. The authors regressed each of the seven metrics separately onto several factors including an indicator for university involvement and a variable for the amount of additional funding other than from the SBIR. In all seven models, the parameter estimate on the university involvement indicator was positive and relatively large in magnitude. However, the estimates were not statistically significant in the models for new employees or trademarks applied for. None of the models had a significant parameter estimate on the additional funding variable, and the signs of the estimates were not consistent.

Link and Scott (2010) provided further evidence to the discussion surrounding successful commercialization of SBIR-funded projects. Link and Scott evaluated the probability of commercialization at the five largest SBIR participating agencies using a binary dependent variable of sales. The binary indicator registers a value of one if there had been some sort of commercialization such as sales of products, processes, services, rights to technology, or spin-off companies. They found that the average of the predicted probabilities of commercialization, by the firms sampled, at each agency, was slightly less than 50 percent.

The SBIR program, which was initially passed in 1982 and renewed several times, provides an opportunity for small businesses to conduct R&D projects that they would not be able to take on without the funding. Small businesses are crucial in bringing innovative technologies to market, however, often lack the funding required to do so. The SBIR program makes it possible for many small businesses to pursue R&D projects that result in a new technology that generates social benefits. In total, the social benefit generated from the SBIR program is much greater than the cost. The importance of the SBIR program in stimulating R&D among small businesses is confirmed through the continued renewal of the program. Much of the literature examining the SBIR program has largely focused on defining and measuring the success of the SBIR program as it relates to outcomes of the program's stated objectives. However, there is a very limited body of research that examines the factors associated with publicly funded project failure; this book begins to fill that gap.

4.4 SBIR and the Department of Energy

The DOE was officially formed in1978 through the amalgamation of several federal energy-related programs, but with roots dating back to World War II and the Manhatten project (see Chap. 3 for a thorough discussion of the history of the DOE). The DOE has a history of advancing scientific innovation to tackle complex challenges that face the US and in some cases the world (DOE-National Labs, n.d.). As such, the DOE has a sizable extramural research budget to facilitate these activities (DOE-OS-SBIR, n.d.). Due to the size of DOE's research budget, the DOE is required to set aside a percentage of its extramural research budget specifically for the SBIR program (DOE-SBIR/STTR, 2022) and awards approximately 360 Phase I and 180 Phase II awards each year through its participating program offices.

There are 13 program offices at the DOE that collaborate with the SBIR program office; Table 4.2 provides a list of these program offices. Through these program offices, research topics are formulated that advance the individual program office's research needs and support the DOE's overall mission. Of these research topics,

Table 4.2 DOE-SBIR participating program offices

Fossil Energy
Fusion Energy Sciences
Nuclear Energy
Energy Efficiency & Renewable Energy
Advanced Scientific & Computing Research
Basic Energy Sciences
Biological and Environmental Research
High Energy Physics
Nuclear Physics
Electricity
Cybersecurity, Energy Security & Emergency Response
Environmental Management
Defense Nuclear Nonproliferation R&D

Sourced from: https://science.osti.gov/sbir/Research-Areas-and-Impact

more than 60 topics and 250 subtopics are available as eligible research avenues for certain businesses to receive an SBIR to support their relevant R&D work (DOE-SBIR/STTR, n.d.).

4.5 Conclusion

In conclusion, the SBIR program is an important source of funding for small entrepreneurial firms conducting R&D intended to advance their technologies to a commercializable product or service. There have been many studies that evaluate the various facets of the program ranging from policy evaluations and employment growth to measuring social benefit of the program. Nonetheless, it is generally agreed that the SBIR program is important in perpetuating the development of new technologies by small firms given its scale in the United States. Therefore, it is central to the analysis performed in this monograph since it provides a relevant real-world case study to evaluate the probability that a firm engaged in an R&D project experiences failure.

The SBIR program is administered by government agencies that have a large research budget. These agencies are required to set aside a percentage of their research budget specifically for their own administration of the SBIR program. The DOE is one of several US government agencies that participate in the SBIR program, and the SBIR program helps fill a critical role for the DOE to create new technologies through innovation. This is accomplished through the DOE's 13 program offices that provide funding to support their high-priority research needs. As such, the data used in the empirical study discussed later come from Phase II awards given through the DOE's SBIR program.

Chapter 5
A Theoretical Model of R&D Project Success or Failure

Abstract This chapter presents a new-to-the-literature microeconomic theoretical model that describes whether a research project succeeds or fails. The model is cast in terms of the success or failure of the search process for a commercializable technology where a failure is defined as negative expected profits. The model considers success and failure to be perfect complements, and principal investigators (PIs) are defined as key input for the firm. The revenue and costs of a new technology are used to create the link between research and development (R&D) inputs and expected revenue from a new technology. Using the expected revenue and costs of the new technology, the profit-maximizing choice of R&D inputs is formulated, and the cases for success or failure are described. The model in this chapter can be considered a structural form model that informs the empirical model described in later chapters.

5.1 Introduction

In the following chapter, a theoretical framework for thinking about how the characteristics of a key individual, in a firm, involved in the research and development (R&D) process may be associated with project failure is proposed; such a construct is new to the literature.

Failure, in general, is often defined in terms of the lack of success of an undertaking, and there may be varying degrees of failure coupled with varying degrees of success. Failure can also be thought of as a binary descriptor to describe the situation when the initial undertaking was not fully successful in terms of the original plan. A project, in this setting, refers to the R&D process a firm engages in to produce a new commercializable technology. The model presented here considers the success and failure of an R&D project to be perfect complements, that is a binary indicator, and principal investigators (PIs) are defined as key individuals within the firm.

M. Boyce, *Principal Investigators and R&D Failure*, Academic Entrepreneurship, Scientists, and Scientific Careers, https://doi.org/10.1007/978-3-031-43608-6_5

When a firm, along with the PI, undertakes an R&D project, there is uncertainty about the success of the project as well as about any potential revenues if the project leads to a commercialized technology that enters the market as an innovation. Uncertainty results from many factors; in this chapter, the focus is on a particular factor, namely the ability of the project PI. The firm's PI is one, within the R&D process, that has the ultimate decision-making power to drive the direction of the project. We posit that experience, E, of the PI is related to the success or failure of an R&D project—more experience is associated with a greater likelihood of success and less experience is thus associated with a greater likelihood of failure.

The PI, in concert with the firm, forms a team, and ultimately a new technology is chosen to pursue (Leyden & Link, 2015); these actions—the formation of the team and the choice of a technology to pursue—define the R&D project. The experience of the PI is thus reflected in the selection of the team and the project to pursue, and therefore, the experience of the PI and the available resources from the firm are ultimately the factors that will be related to the success of the project pursued. Given the resources from the firm that are available for the selected R&D project, a PI with greater experience should tend to influence the selection of a portfolio of inputs that result in a project with higher likelihood of success relative to a lesser experienced PI.

Choosing the members of the R&D team is one of the most important decisions the PI makes. Without a team, or with a team that has insufficient ability, the PI may not have the necessary human capital to perform the required R&D successfully and/or to bring any resulting technology to market. PIs with more experience may have a larger and more relevant network—an internal to the firm network and an external network—of available team members; a larger network provides the PI with a greater potential supply of human capital to choose from. This point was emphasized by Schott and Sedaghat (2014, p. 472): "... that the size of the network around entrepreneurs positively affects innovation ..." and "The more an entrepreneur networks, overall, the more innovative the entrepreneur is likely to be ...".

5.2 Revenue and Costs of a New Technology

Thus, the overall resource base available to the PI may be thought of as input to the R&D process; let Q represent a measure of the quantity of the R&D-related inputs available to the project. With Q as a measure of input quantity, $p(Q,E)$ represents the probability of a new marketable technology brought to market. Success is defined as a new marketable technology with positive expected profits. It is assumed that the $p(Q,E)$ function is a positive concave function that exhibits diminishing returns, therefore, $\frac{\partial p}{\partial Q} > 0$, $\frac{\partial^2 p}{\partial Q^2} < 0$, $\frac{\partial p}{\partial E} > 0$, and $\frac{\partial^2 p}{\partial E^2} < 0$.[1]

[1] It is possible that eventually Q could become too large, therefore reducing p at the margin, that is, $\partial p/\partial Q < 0$, although this set of Q is not relevant under the assumption that the PI would never construct a project at that input quantity.

Let $R(q)$ represent a concave function that represents a mapping of the quality of the resulting commercializable technology, that is the quality of the resulting innovation (i.e., the technology brought to market), and q represent the revenues received by the firm that are generated from the sale of the innovation. $R(q)$ exhibits diminishing marginal returns, therefore, $\frac{\partial R}{\partial q} > 0$ and $\frac{\partial^2 R}{\partial q^2} < 0$. The quality of the resulting innovation is assumed to be a concave function of the quantity of the R&D inputs and PI input to the project, $q(Q, E)$,such that $\frac{\partial q}{\partial Q} > 0$ and $\frac{\partial q}{\partial E} > 0$.

The quality of the innovation is a measure of how much consumer surplus the innovation provides to those who purchase it. For example, an innovation that costs less relative to a substitute product has a higher quality (q) than that of the substitute, since consumers would prefer to pay less when choosing between comparable products. Likewise, a product that has superior features that provides greater utility, sufficiently high to compensate for a possible higher price, is another example of higher quality. Thus, the expected revenues, R^e, from the innovation will be:

$$R^e = p(Q,E) * R(q(Q,E)), \tag{5.1}$$

and expected marginal revenues, with respect to either Q or E, will be:

$$MR^e = p'(Q,E) * R(q(Q,E)) + p(Q,E) * R'(q(Q,E)) * q'(Q,E). \tag{5.2}$$

Naturally, there is cost associated with the R&D process. Leyden and Link (2015) describe the entrepreneurial process as a two-step iterative process that incurs cost at each step. They posit that costs are funded by both personal endowments and capital markets depending on the phase of the process. Following Audretsch et al. (2012), costs may be intellectual or physical and are both fixed and variable. In this construct, greater Q increases total cost, at an increasing rate, thus cost is a function of inputs. The cost function can be written as:

$$c = c(Q) \supset c(0) > 0, c'(Q) > 0, c''(Q) > 0. \tag{5.3}$$

5.3 Profit-Maximizing Choice of R&D Inputs

The firm looking to maximize profits will choose an optimal combination of inputs, Q^*, such that it maximizes expected profit:

$$\pi^e(Q,E) = p(Q,E) * R(q(Q,E)) - c(Q). \tag{5.4}$$

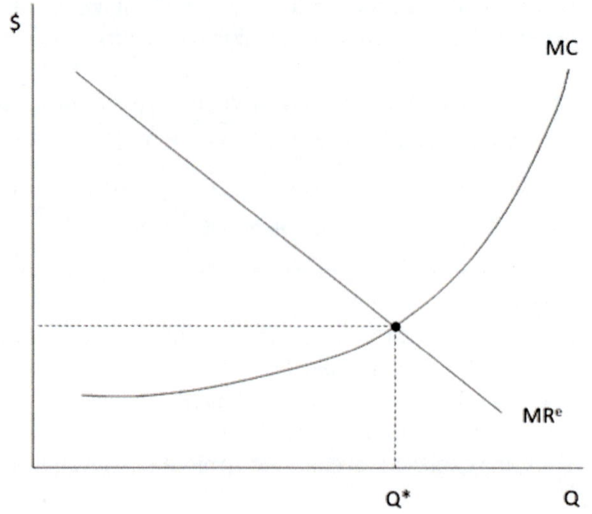

Fig. 5.1 Profit maximizing choice of experience

The optimal quantity of inputs conditional on the experience of the PI, Q^*, is achieved by equating expected the marginal revenue, MR^e, to the marginal cost, MC, of the R&D project (Fig. 5.1):

$$p(Q,E) * R'(q(Q,E)) * q'(Q,E) + p'(Q,E) * R(q(Q,E)) = c'(Q). \quad (5.5)$$

As mentioned previously, at the onset of an R&D project, there is uncertainty around the ultimate success of the project. The PI's experience determines the ability of the quantity of inputs, Q, to generate a given quality, q, of the innovation, and ultimately the maximum expected profit. To illustrate this point, Fig. 5.2 presents the expected profit curves as a function of project quality for two alternative projects, and Fig. 5.3 presents the densities of the project's expected profits. From the two figures, it is clear that the project with PI experience, E_2, (project 2) has profits $\pi_2^e(Q,E_2)$ that are negative even at the profit-maximizing choice of quality. The project with expected profits $\pi_1^e(Q,E_1)$ (project 1) has positive expected profits at the optimal quality level.

Project 2 has negative expected profits because the quantity of inputs, Q^*, that the PI is able to transform into a marketable product can only be sold at a price less than the average cost. Defining negative expected profits as the indicator of failure, $\pi^e(Q^*, E) \geq 0$, must hold true for the project to be categorized as a success. Therefore, project 1 would be one worth pursuing and project 2 would not.

From this conclusion, it can be inferred, ceteris paribus, that the PI associated with project 1 has greater experience than the PI associated with project 2. Thus, this model suggests that the experience base of the R&D project's PI is positively related to project success. It follows then that the experience base of the R&D project's PI is negatively related to project failure.

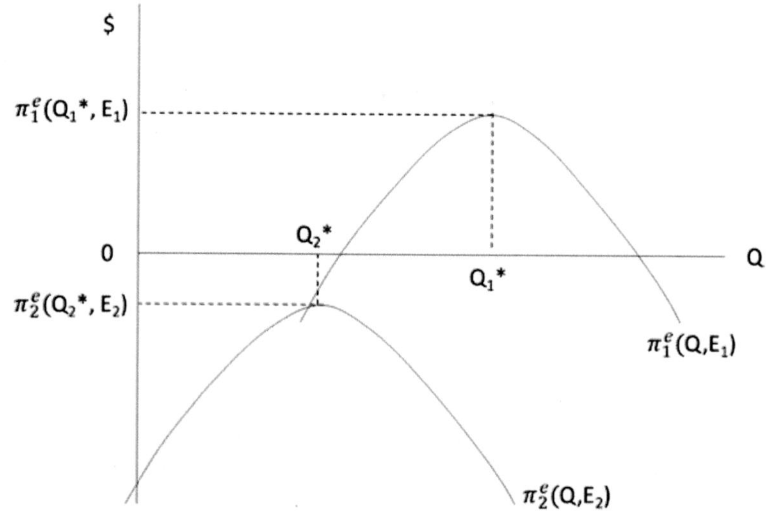

Fig. 5.2 Profit maximization success and failure

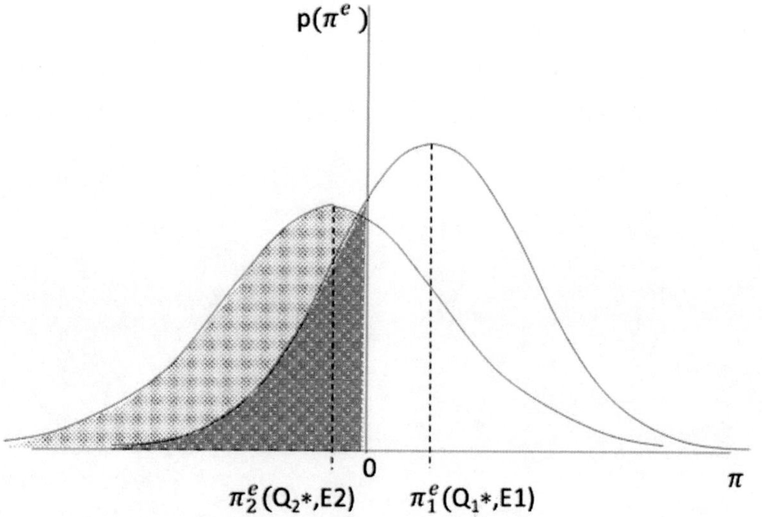

Fig. 5.3 Expected profit distributions

Chapter 6
Department of Energy and SBIR Data

Abstract The data used to examine the testable hypotheses that were formulated through the theoretical model are discussed in this chapter. There is also a discussion that details the legislation that required the data to be collected. Specifically, the National Academy of Sciences (the Academy) was legislated to evaluate the Small Business Innovation Research (SBIR) program and to provide their findings to Congress. As part of their evaluation, they conducted a survey of SBIR program participants, of which the full population of the Department of Energy (DOE) SBIR program participants for the survey period were surveyed. Using the survey data from the Academy's survey of the DOE SBIR program, two dependent variables (Failure and Technical Failure) are defined for use in the empirical analysis. Descriptive statistics are also provided for the dependent variables.

6.1 Background

Small businesses contribute substantially to the US economy. As of 2014 (the most recent year the complete data are available), small business[1] share of private nonfarm gross domestic product was approximately 43.5 percent of the total, and as of 2014, it contributed to just under half of the total number of employees in the United States in terms of nonfarm payrolls (Kobe & Schwinn, 2018). Of this important sector in the economy, the "Small Business Innovation Research (SBIR) program remains the nation's largest innovation program for small business" (National Academies, 2016, p. 1). Hence, the data used in this analysis are information gathered from recipients of a Phase II SBIR award. Specifically, the data are a random sample of 225 Phase II projects (referred to later as the full sample) funded by the Department of Energy's (DOE's) SBIR program in 2001 through 2010.

[1] Small business in this case is defined as having fewer than 500 employees. According to the Small Business Administration size standard, most small businesses qualify as small businesses under this rule. See https://www.sba.gov/sites/default/files/files/Size_Standards_Table.pdf

M. Boyce, *Principal Investigators and R&D Failure*, Academic Entrepreneurship, Scientists, and Scientific Careers, https://doi.org/10.1007/978-3-031-43608-6_6

The data were collected in 2014 through a survey (second-round survey) conducted by the Academy. The Academy is a "private, nonprofit institutions that provide expert advice on some of the most pressing challenges facing the nation and the world" (Academy, n.d.). As a result of legislation written in the SBIR reauthorization of 2000 (Public Law 106-554) and further emphasized by the 2011 SBIR program reauthorization (S. 493), the Academy was tasked with evaluating the SBIR program. The reauthorization that occurred in 2000 states that,

> … each Federal agency with a budget of more than $50 million for its SBIR Program for FY 1999 to enter into an agreement with the National Academy of Sciences for the National Research Council to: (1) conduct a study of the value and benefits achieved by the Program; and (2) make appropriate recommendations for Program improvement (Sec 108).

The research conducted as a result of the reauthorization of 2000, which included a similar survey (first-round) as the second-round survey, led to a series of Academies reports published from 2004 to 2009. These reports focused on how the SBIR program contributed to technological innovation in the United States at the largest five federal agencies (based on SBIR budgets), which included the DOE (Academy, 2016). The Academies were also tasked with providing recommendations on how to improve the SBIR program.

When the Senate reauthorized the SBIR program in 2011, the legislation reemphasized the requirement to evaluate the program by amending:

> … the Small Business Reauthorization Act of 2000 to continue NAS [National Academies of Sciences] evaluation of the SBIR program, as well as reports on such evaluation from the National Research Council to participating agency heads and the small business committees (Sec. 307).

Therefore, the Academies conducted a second-round assessment that included a survey in 2014 that was "… sent to every PI who received a Phase II award from DOE, FY 2001–2010" (National Academies, 2016, p. 3). The reports from the second round survey, while specific to each agency's needs, were intended to:

> 1. Review institutional initiatives and structural elements contributing to programmatic success, including gap funding mechanisms such as applying Phase II-plus awards more broadly to address agency needs and operations and streamlining the application process. 2. Explore methods to encourage the participation of minorities and women in SBIR and STTR. 3. Identify best practice in university-industry partnering and synergies with the two programs. 4. Document the role of complementary state and federal programs. 5. Assess the efficacy of post-award commercialization programs. (Academy Assessment, n.d.)

6.2 Survey Data

The raw survey data from the Academies 2014 survey provide information on the population of 244 Phase II DOE SBIR awards between the 2001 and 2010. However, the completeness of the survey responses was not consistent across each award. Of the 244 survey responses received, 19 random awardees did not disclose an outcome/status of the funded project, so those projects were excluded from the population. Therefore, this book draws on a random sample of 225 responses, of which 170 were from a principal investigator (PI), to the second-round survey to explore the role PIs play in project failure. Table 6.1 shows the distribution of awards by year for both the population and the random sample. As can be seen in the table, the samples of awards are distributed similarly to the population.

Additionally, Table 6.2 shows the distribution of outcomes of the funded projects. Projects that successfully delivered their product/process/service (technology) and were in use by their target population occurred at the highest rate of nearly one out of four project outcomes. Another 4.1% of projects resulted in their technology being used by an unexpected population of users. These two outcomes can be considered as success outcomes. The second most frequent status at the time of the survey, which was noted by 23% of the firms that answered the question, was that the project was continuing post-award. In terms of the SBIR program, this could likely be referred to as Phase III of the program given Phase III is not an SBIR-funded phase. Two other responses indicate the project was ongoing at the time of the survey, specifically commercialization being in progress (12.3%) and the project persisting in Phase II of the SBIR program (2%). Therefore, at the time of the survey, 37.3% of the projects were still in progress. One project status that may not represent a binary measure of success/failure is the case where efforts at the firm ended, but the project resulted in sales, licenses, or additional funding. This status was noted by 7.8% of the firms that answered the question and implies that there is not an active market or use for the technology developed. Finally, firms that are no longer continuing and the project did not result in sales or licensing can be considered as a failure outcome. Forty-five or 18.4% of the 244 DOE SBIR Phase II award recipients experienced failure, according to this measure.

Table 6.1 Distribution of awards by year

Award year	Count of awards (Population)	Count of awards (Sample)	Share of awards (Population)	Share of awards (Sample)
2001	13	11	5.33%	4.89%
2002	8	8	3.28%	3.56%
2003	14	13	5.74%	5.78%
2004	18	18	7.38%	8.00%
2005	14	13	5.74%	5.78%
2006	21	18	8.61%	8.00%
2007	29	26	11.89%	11.56%
2008	35	30	14.34%	13.33%
2009	41	38	16.80%	16.89%
2010	51	50	20.90%	22.22%
Total	244	225	100%	100%

Table 6.2 Distribution of project outcomes

Status of project at time of survey	Count	Share
Commercialization is underway	30	12.3%
Efforts at this company have been discontinued no sales or additional funding resulted from this project	45	18.4%
Efforts at this company have been discontinued the project did result in sales, licensing of technology, or additional funding	19	7.8%
Products/processes/services are in use by population/customer/consumers not anticipated at the time of the award (for example, in a different industry)	10	4.1%
Products/processes/services are in use by target population/customer/consumers	60	24.6%
Project has not yet completed SBIR/Small Business Technology Transfer (STTR)-funded research	5	2.0%
Project is continuing post-award technology development	56	23.0%
No response	19	7.8%
Total	244	100%

6.3 Measures of Failure

In the empirical analysis presented later, two separate binary measures of failure are defined. As mentioned previously, failure may not always be binary; however, the data used for this monograph provides the opportunity to look at failure as a binary measure as opposed to others. The first measure captures a general measure of failure and is defined as a Phase II–funded project that was discontinued with no sales from the developed technology or any additional funding (*Failure*). *Failure* may occur due to any of the following issues:

1. The level of technical risk was too high.
2. The principal investigator left.
3. Technical failure or difficulties.
4. Market demand was too small.
5. The firm shifted priorities.
6. Inadequate sales capability.

As shown in Table 6.2, there were 19 randomly missing responses to the survey question regarding project status or outcome. Responses to that question are the basis for the dependent variable(s) analyzed later. Therefore, the random sample size of DOE SBIR Phase II–funded projects considered here is 225. Of the 225 DOE Phase II–funded projects studied here, 45 failed, which translates to a 20 percent failure rate. A similar measure of failure was used by Link and Wright (2015) who used a random sample of 1878 Phase II projects, across five agencies, from the first-round survey to estimate the probability an SBIR-funded project fails. Of the 1878 Phase II projects studied by Link and Wright (2015), 624 failed, representing a 33 percent failure rate.[2] Likewise, Andersen et al. (2017) define a similar measure

[2]There are 154 DOE Phase II projects in the 1878 sample. Although the failure rate by agency is not provided by Link and Wright (2015), if available it would have allowed a comparison to the

Table 6.3 Dependent variables descriptive statistics ($n = 225$)

Variable	Mean	StdDev	Range
Failure	0.200	0.401	0/1
TechFailure	0.062	0.242	0/1

of failure and used a random sample of 461 SBIR projects funded by the National Institute of Health (NIH), from 1992 through 2001, to estimate the probability a project fails. The failure rate of the 461 NIH-funded projects was found to be 21.5 percent. Thus, the failure rate identified in the data for this analysis is not out of line with two of the studies that are most relevant to the scope of this book and the analysis that follows.

The second dependent variable is a narrower measure of failure, namely failure for technical reasons (*TechFailure*). Specifically, ones that failed for technical reasons or because technical risk was too high are defined as failing for technical reasons. This more specific measure of project failure is considered in this piece because having a narrower understanding of reasons that a project may fail can help policymakers develop more pointed policies that may be more effective for guiding firms that are susceptible to a specific type of failure. Of the 45 projects that failed for any reason, 14 or 31 percent of those failed for technical reasons. This book is the first study known to analyze covariates with SBIR project technical failure.

Further, it is assumed that reducing the probability of failure is an objective of the SBIR program. The SBIR program has limited funds to allocate, therefore program administrators should want to allocate funds as efficiently as possible to maximize the program's social return on investment. When appropriating funds to a firm, the SBIR expects the firm to succeed and generate some social benefit, else the award would not be given. Nonetheless, this is not a strong assumption since it is clear the SBIR program would certainly prefer a positive social return on the awards granted over the alternative (Table 6.3).

6.4 Conclusion

This chapter introduced the data used in the analysis that follows. Specifically, the data are the responses to a 2014 survey from the National Academies sent to all PIs who received a Phase II SBIR award from the DOE between 2001 and 2010. Therefore, the initial data represent the population of Phase II award recipients and provide a rich information set that allows the theoretical model presented in Chap. 5 to be tested empirically. Due to random survey recipients failing to provide a response regarding the status or outcome of their funded project, a random sample of 225 firms is used as the initial sample to define the dependent variables of project failure.

DOE sample studied here.

Chapter 7 uses these data to estimate a probit model of project failure using firm and project characteristics captured in the Academies 2014 survey. Chapter 8 extends the analysis performed in Chap. 7 by augmenting the specifications in Chap. 7 with information specific to PIs using the same survey data.

Chapter 7
Collaborative R&D Firm and Project Characteristics: Empirical Insights

Abstract In this chapter, the reduced form specification of the theoretical model including the empirical data used to estimate the model is described. The explanatory variables are motivated by past studies and are mostly firm or project characteristic variables (firm size, industry, experience, principal investigator (PI)/owner's gender). Descriptive statistics for the independent variables are discussed, and initial expectations for the direction of the relationships between the explanatory variables and the two measures of failure are provided. Results from the probit estimations of various specifications are provided and, statistically significant factors are discussed. This chapter also serves the purpose of confirming, at least directionally, findings from two similar past studies. For example, the estimated relationship between project failure and having a female principal investigator or a female firm owner is negative, which is directionally consistent with the results found in Link and Wright (2015) and Andersen et al. (2017). However, the estimates produced from the data in this study are not statistically significant. Additionally, the model estimates show statistically significant negative relationships between project failure and having relevant experience and increased human capital.

7.1 Introduction

This chapter begins with an explanation of the empirical model used to test the theoretical model laid out in Chap. 5 with the available data from the Department of Energy (DOE) Small Business Innovation Research (SBIR) program. The set of data used in this analysis represent firm and project characteristics such as past experience, size of the firm, and industry among others. Following this is an explanation of the variables considered in the analysis that are motivated by two relevant studies with hypotheses of the expected relationship with failure. The analysis in this chapter is considered the initial analysis because the intent of this chapter is largely focused on confirming the results of the two studies that motivated the initial variable selection.

M. Boyce, *Principal Investigators and R&D Failure*, Academic Entrepreneurship, Scientists, and Scientific Careers, https://doi.org/10.1007/978-3-031-43608-6_7

Descriptive statistics are provided and discussed followed by an analysis of the independent variables and their relationship between *Failure* and then *TechFailure* by segmenting projects by those that failed and those that did not. Through the segmentation, preliminary inspections of the hypotheses mentioned above are discussed.

Following the analysis of summary statistics, three sets of models are presented based on the execution of the empirical formulation described earlier. Each set of estimations provides results for both *Failure* and *TechFailure*. The first set of models contains the fewest explanatory variables and is used as the base set, which was motivated by Link and Wright (2015) and Andersen et al. (2017). Following the first model set, two additional sets of estimation results are provided, whereas each additional model set introduces another firm or project characteristic explanatory variable. The models are estimated in the manner described to ensure interactions between the independent variables do not impact the prior set's parameter estimates in any meaningful way. As such, the chapter concludes with a presentation of variable inflation factors to further ensure interactions between the explanatory variables are not statistically significant.

7.2 Initial Probit Model

According to the structural model laid out in Chap. 5, the experience, *E*, of the PI impacts the probability of bringing new technology to market, the expected revenues a firm can earn as a function of *q* (quality of the innovation), and the cost associated with the project. However, due to the data available, a reduced form specification is used where project failure is a binary outcome as opposed to the continuous formulation using expected profits as the measure of failure. Thus, research and development (R&D) inputs, *Q*, and experience, *E*, are used to estimate the probability of project failure as a binary model. Given the earlier work by Link and Wright (2015) and Andersen et al. (2017), and the similarities in the data they used with those used in this book, the first model presented looks to those studies to identify variables in the data that represent *Q* and *E* and may be associated with project failure. That is, a simple probability model of failure is employed to identify sets of covariates that are associated with SBIR project failure. The model is defined as:

$$failed = I\left(X_i\beta + \epsilon_i > 0\right), \tag{7.1}$$

where *failed* (*Failure*, *TechFailure*) is a binary variable, *I* is an indicator function, **X** is a vector of firm and project characteristics, and $\epsilon \sim N(0, 1)$. Multiple constructs of **X** are also used to analyze the associations among different firm and project characteristics with *Failure* and *TechFailure*.

7.2.1 Independent Variables Considered

Three independent variables that were used in Link and Wright (2015) are available in the DOE dataset and are used in this analysis. In addition, three of the independent variables used in Andersen et al. (2017) are also used in this analysis. These variables as well as additional variables unique to this empirical work that are used in the initial estimations of Eq. 7.1 are defined in Table 7.1. The definitions of the two dependent variables, *Failure* and *TechFailure*, are also provided below.

The first variable considered, *FemalePI*¸ is hypothesized, based on the literature, to have a negative relationship with *Failure*. If a firm has a female PI, then it is expected that the project is less likely to fail, on average, than a firm with a male PI; this relationship has been shown to be statistically significant by Link and Wright (2015). Inherent differences in the nature of males versus females, such as a female PI tending to be "more innovative and critical thinking in problem solving" (Link & Wright, 2015, p. 445) may provide context to this result. Similarly, Andersen et al. (2017) considered the gender of the owner as opposed to the gender of the PI in their study and found a statistically significant relationship between SBIR project failure and female ownership. They also found this relationship to be statistically significant even when controlling for possible endogeneity of the variable. Given the findings from Andersen et al. (2017), *FemaleOwner* is used as an alternative measure to *FemalePI* in conjunction with the other covariates specified below. Thus, based on the literature, the hypothesized relationship of *FemaleOwner* with *Failure* is negative.

No hypothesis is offered for the variable *Employees*, which is a measure of firm size. On the one hand, larger firms might take on more complex projects. As a project becomes more complex, the probability it will fail might increase. On the other

Table 7.1 Variable definitions

Variable	Definition
Failure	= 1 if project was discontinued with no sales or additional funding received; 0 otherwise
TechFailure	= 1 if failure = 1 and the reason for failure was of a technical nature; 0 otherwise
FemalePI	= 1 if the principal investigator is a female; 0 otherwise
FemaleOwner	= 1 if the owner of the firm is a female; 0 otherwise
Employees	Number of employees in the firm at the time the phase II award was received
SimAwardsDummy	= 1 if received phase II award for a similar technology previously; 0 otherwise
ProfInvolved	= 1 if university faculty worked on the project; 0 otherwise
DEE	= 1 if project was in the energy or environmental sector only
DEng	= 1 if project was in the engineering sector only
Other	= 1 if project was in only one sector and not in DEE or DEng
DMT	= 1 if project was in multiple sectors

hand, larger firms might have more resources to draw on in the event that unexpected events occur during the Phase II research. Thus, in this case, the probability that a project will fail decreases with firm size.

SimAwardsDummy is a binary variable that indicates whether any prior similar awards have been received and is expected to have a negative relationship with *Failure*, as shown in prior studies. *SimAwardsDummy* is a proxy for research experience, specifically, experience in researching a related technology. Having at least some past research experience is expected, on average, to reduce the likelihood that the firm's project would fail.

ProfInvolved is a binary indicator for whether university faculty were involved in the project. This variable is used as an additional measure of human capital. Having university faculty involved in the project likely brings some level of expertise or experience that should increase the level of human capital involved in the project. Greater human capital, *ceterus paribus,* should reduce the probability of project failure. Thus, *ProfInvolved* is hypothesized, based on the literature, to have a negative relationship with *Failure*.

Moving beyond the variables suggested by the literature, fixed effect controls for the SBIR projects technology sector are also considered. The data indicate that for some firms sampled, the technology currently being funded falls into multiple technology sectors; these projects are captured by a dummy variable *DMT* for projects in multiple technology sectors. Thus, as a measure of project complexity, *DMT* is hypothesized, ceteris paribus, to be positively associated with SBIR project failure.

Further, the relationship between each of the previously mentioned variables and failure is expected to have the same directional relationship with both *Failure* and *TechFailure*, with the exception of *Employees.* Since no hypothesis is proposed for *Employees*, no speculation is offered as to whether the relationship between *Employees* and *Failure*, and, *Employees* and *TechFailure,* should be the same.

7.2.2 Descriptive Statistics

Descriptive statistics for the variables used in Eq. (7.1) are presented in Table 7.2. The number of observations used is 169 of the 225 Phase II projects due to some firm's nonresponse to one or more of the survey questions used to calculate the independent variables. The mean value of *Failure* in this sample is 0.213 compared to 0.20 in the full sample of 225 projects (see Table 6.1). Thus, 21.3 percent of the projects in the sample failed. The mean value of *TechFailure* in this sample is 0.071 compared to 0.062 in the full sample (see Table 6.1), thus 7.1 percent of the projects in the sample failed for technical reasons. The mean number of projects with a *FemalePI* is 5.3 percent, and the mean number of projects with a *FemaleOwner* is 5.9 percent. The number of *Employees* within a firm at the time the Phase II project was received is, on average, just over 36. The share of firms that had at least one past award in a similar technology, *SimAwardsDummy*, is 60.4 percent. The mean number of projects with a *ProfInvolved* is 0.29. The energy and environment sector

Table 7.2 Descriptive statistics on variables used in Eq. (7.1)

Variable	Mean	StdDev	Range
Failure	0.213	0.411	0/1
TechFailure	0.071	0.258	0/1
FemalePI	0.053	0.225	0/1
FemaleOwner	0.059	0.237	0/1
Employees	36.124	49.106	1–300
SimAwardsDummy	0.604	0.491	0/1
ProfInvolved	0.290	0.455	0/1
DEE	0.207	0.406	0/1
Deng	0.142	0.350	0/1
Other	0.166	0.373	0/1
DMT	0.485	0.501	0/1

(*DEE*) accounts for 20.7 percent of the projects in the sample, the share of projects in the engineering sector (*DEng*) is 14.2 percent, projects in one sector other than energy and environment or engineering (*Other*) account for 16.6 percent, and projects in multiple technology sectors (*DMT*) represent 48.5 percent of the sample.

7.2.3 *Relationships Between Independent Variables and Failure*

To explore the relationship among the independent variables and *Failure* using the model in Eq. (7.1), descriptive statistics on the variables being considered were examined. Specifically, the 169 Phase II projects were segmented into those that failed ($n = 33$) and those that did not fail ($n = 133$) based on the variable *Failure*. As shown in Table 7.3, 2.8 percent of the projects that failed had a female PI compared to 6 percent of the projects that did not fail. Similarly, 2.8 percent of the projects that failed had a female owner compared to 6.8 percent of the projects that did not fail. The mean number of employees in projects that failed is 40, which is relatively close to the mean number of employees in the projects that did not fail, namely 35. Considering the binary indicator of having received a past similar award in a related technology field, *SimAwardsDummy*, about 31 percent of the projects that failed had at least one previous award compared to about 68 percent of the projects that did not fail. Further, of the projects that experienced *Failure*, approximately 11 percent had university faculty involved, compared to nearly 34 percent of firms with *ProfInvolved* that did not experience *Failure*. These initial findings are consistent with my original hypotheses.

Moving to the technology sector fixed effects, projects that failed and that are categorized into a single sector, the *DEE*, *DEng*, and *Other*, account for 3.0, 3.6 and, 5.9 percent of the sampled projects, respectively. For the projects that did not fail, the *DEE*, *DEng*, and *Other* represent 17.8, 10.6, and 10.6 percent of the sample,

Table 7.3 Independent variables grouped by *Failure*

Variable	Mean	StdDev	Range
***Failure*: 1 (*n* = 36)**			
TechFailure	0.333	0.478	0/1
FemalePI	0.028	0.167	0/1
FemaleOwner	0.028	0.167	0/1
Employees	40.083	60.259	1–300
SimAwardsDummy	0.306	0.467	0/1
ProfInvolved	0.111	0.319	0/1
DEE[a]	0.030	0.351	0/1
DEng[a]	0.036	0.378	0/1
Other[a]	0.059	0.454	0/1
DMT[a]	0.089	0.500	0/1
***Failure*: 0 (*n* = 133)**			
TechFailure	0.000	0.000	0–0
FemalePI	0.060	0.239	0/1
FemaleOwner	0.068	0.252	0/1
Employees	35.053	45.835	1–224
SimAwardsDummy	0.684	0.467	0/1
ProfInvolved	0.338	0.475	0/1
DEE[a]	0.178	0.419	0/1
DEng[a]	0.106	0.343	0/1
Other[a]	0.106	0.343	0/1
DMT[a]	0.396	0.502	0/1

[a]Since these are fixed effects controls, *n* = 169 so that the means represent the share of each sector in each group relative to the total sample as opposed to the within-group share

respectively. Projects that are classified in multiple sectors and failed represent 8.9 percent of the projects sampled, and those that did not fail account for 39.6 percent. Thus, the share of projects classified into multiple technology sectors that did not fail is much greater than its share of projects that did fail, which does not support the original hypothesis of this variable's expected association with *Failure*.

7.2.4 Relationships Between Independent Variables and *TechFailure*

To explore the relationship among the independent variables and *TechFailure* using the model in Eq. (7.1), the 169 Phase II projects were segmented into those that failed for a technical reason (*n* = 12) and those that did not fail (*n* = 157). As shown in Table 7.4, none of the projects that failed for technical reasons had a female PI compared to 5.7 percent of the projects that did not fail for technical reasons. Similarly, none of the projects that failed for technical reasons had a female owner,

Table 7.4 Independent variables grouped by *TechFailure*

Variable	Mean	StdDev	Range
***TechFailure*: 1 (*n* = 12)**			
FemalePI	0.000	0.000	0/1
FemaleOwner	0.000	0.000	0/1
Employees	26.083	14.286	5–56
SimAwardsDummy	0.250	0.452	0/1
ProfInvolved	0.083	0.289	0/1
DEE[a]	0.012	0.389	0/1
DEng[a]	0.000	0.000	0–0
Other[a]	0.030	0.515	0/1
DMT[a]	0.030	0.515	0/1
***TechFailure*: 0 (*n* = 157)**			
FemalePI	0.057	0.233	0/1
FemaleOwner	0.064	0.245	0/1
Employees	36.892	50.736	1–300
SimAwardsDummy	0.631	0.484	0/1
ProfInvolved	0.306	0.462	0/1
DEE[a]	0.195	0.409	0/1
DEng[a]	0.142	0.361	0/1
Other[a]	0.136	0.355	0/1
DMT[a]	0.456	0.501	0/1

[a]Since these are fixed effects controls, *n* = 169, so the means represent the share of each sector in each group relative to the total sample as opposed to the within-group share

while 6.4 percent of the projects that did not fail for technical reasons had a female owner. This finding is consistent with the earlier findings for the *FemalePI* and *FemaleOwner* indicators grouped by *Failure*.

The mean number of employees in projects that failed for technical reasons is just greater than 26 compared with the mean number of employees in the projects that did not fail for technical reasons of nearly 37. Therefore, there are 11 less *Employees,* on average, who worked on a project that experienced *TechFailure* than those that did not. This association is the opposite of the relationship between *Employees* and *Failure*, which resulted in about five more *Employees*, on average, that experienced *Failure* than not. This result, perhaps, implies a nonlinear relationship between *Employees* and projects that fail, for any reason as will be discussed below.

Considering past experience, as measured by *SimAwardsDummy*, 25 percent of projects that failed for technical reasons received a past award, while approximately 63 percent of projects that did not fail for technical reasons received a past award. These findings are consistent with the earlier findings of *SimAwards* segmented by *Failure*.

Of the firms that experienced *TechFailure*, approximately 8 percent had university faculty working on the project, compared to nearly 31 percent of firms with

ProfInvolved that did not experience *TechFailure*. These findings are consistent with the earlier findings of *ProfInvolved* segmented by *Failure.*

Moving to the technology sector fixed effects, projects that failed for technical reasons and are categorized into a single sector, the *DEE, DEng*, and *Other* groups, account for 1.2, 0.0, and 3.0 percent of the sampled projects, respectively. For the projects that did not fail for technical reasons, the *DEE*, *DEng*, and *Other* represent 19.5, 14.2, and 13.6 percent of the sample, respectively. Further, projects that were classified in multiple sectors and experienced *TechFailure* represent 3.0 percent of the projects sampled and those that did not fail for technical reasons account for 45.6 percent. Thus, the share of projects classified into multiple technology sectors that did not experience technical failure is much greater than its share of projects that did experience technical failure, similar to the *DMT* results when grouped by *Failure.*

7.3 Initial Probit Model Results

Parameter estimates and marginal effects from the estimation of Eq. (7.1) using different project and firm characteristics are shown in Tables 7.5, 7.6, 7.7, and 7.8. Each of the tables present findings from three separate models: two models with *Failure* as the response and one model with *TechFailure* as the response. The two models for *Failure* have the same specifications except either *FemalePI* or *FemaleOwner* is used in the estimation. Further, the random sample size for all models presented here forward is 169.

7.3.1 *Set One Model Results*

The variables used in the first estimations of Eq. (7.1), shown in Table 7.5, are those that are suggested by the literature. Focusing first on the models of *Failure,* as shown in Table 7.5, both *FemalePI* and *FemaleOwner* parameter estimates support their hypothesized relationships with *Failure*. That is, both *FemalePI* and *FemaleOwner* parameter estimates are negative, and the two estimates have similar values. However, these estimates are not statistically significant. The coefficients on *Employees* are positive in the two *Failure* models, and when *FemalePI* is included, the coefficient is slightly greater than when *FemaleOwner* is used in the model. Due to the potential nonlinear association between *Failure* and *Employees,* a relationship was not hypothesized; thus, these estimates support the notion that marginally, larger firms are more likely to experience project failure. However, neither of the estimates of *Employees* are statistically significant. The *SimAwardsDummy* variable is used to analyze whether having received any past awards in a related technology is associated with SBIR project failure. The coefficient on *SimAwardsDummy* is negative in the two *Failure* models and both are significant at the 0.001 level. The

Table 7.5 Set one probit results from Eq. (7.1)

	(1)		(2)		(3)	
	Failure	Marginal effects	*Failure*	Marginal effects	*TechFailure*	Marginal effects
FemalePI	−0.569	−0.120			–[a]	–[a]
	(0.568)	(0.0931)				
FemaleOwner			−0.585	−0.123	–[a]	–[a]
			(0.554)	(0.0899)		
Employees	0.00137	0.000357	0.00123	0.000318	−0.00389	−0.000485
	(0.00226)	(0.000585)	(0.00229)	(0.000592)	(0.00219)	(0.000315)
SimAwardsDummy	−0.932***	−0.242***	−0.928***	−0.241***	−0.797*	−0.0994*
	(0.229)	(0.0525)	(0.229)	(0.0526)	(0.313)	(0.0419)
Intercept	−0.342		−0.336		−0.984***	
	(0.178)		(0.180)		(0.200)	
Wald χ^2	17.92**		18.19**		15.49***	
Likelihood ratio	18.15***		18.24***		7.512*	

Robust standard errors in parentheses
*$p < 0.05$ **$p < 0.01$ ***$p < 0.001$
[a]*FemalePI* and *FemaleOwner* are not used in the model due to lack of variation

Table 7.6 Set two probit results from Eq. (7.1)

	(1)		(2)		(3)	
	Failure	Marginal effects	*Failure*	Marginal effects	*TechFailure*	Marginal effects
FemalePI	−0.599	−0.123			–[a]	–[a]
	(0.565)	(0.0896)				
FemaleOwner			−0.609	−0.125	–[a]	–[a]
			(0.544)	(0.0860)		
Employees	0.00224	0.000565	0.00205	0.000516	−0.00394	−0.000485
	(0.00241)	(0.000599)	(0.00244)	(0.000607)	(0.00268)	(0.000368)
SimAwardsDummy	−0.847***	−0.213***	−0.843***	−0.212***	−0.709*	−0.0873*
	(0.232)	(0.0522)	(0.232)	(0.0523)	(0.301)	(0.0387)
ProfInvolved	−0.678*	−0.171*	−0.676*	−0.170*	−0.507	−0.0625
	(0.322)	(0.0778)	(0.320)	(0.0772)	(0.453)	(0.0552)
Intercept	−0.267		−0.261		−0.925***	
	(0.181)		(0.182)		(0.211)	
Wald χ^2	19.22***		19.75***		13.40**	
Likelihood ratio	23.45***		23.53***		8.93*	

Robust standard errors in parentheses
* $p < 0.05$ ** $p < 0.01$ *** $p < 0.001$
[a]*FemalePI* and *FemaleOwner* are not used in the model due to lack of variation

Table 7.7 Set three probit results from Eq. (7.1)

	(1)		(2)		(3)	
	Failure	Marginal effects	Failure	Marginal effects	TechFailure	Marginal effects
FemalePI	−0.601	−0.123			–[a]	–[a]
	(0.559)	(0.0885)				
FemaleOwner			−0.601	−0.123	–[a]	–[a]
			(0.541)	(0.0867)		
Employees	0.00231	0.000582	0.00208	0.000524	−0.00397	−0.000489
	(0.00245)	(0.000607)	(0.00248)	(0.000616)	(0.00280)	(0.000382)
SimAwardsDummy	−0.839***	−0.211***	−0.839***	−0.211***	−0.711*	−0.0875*
	(0.226)	(0.0514)	(0.227)	(0.0516)	(0.296)	(0.0385)
ProfInvolved	−0.673*	−0.170*	−0.673*	−0.169*	−0.511	−0.0628
	(0.322)	(0.0779)	(0.319)	(0.0773)	(0.441)	(0.0539)
DMT	−0.0588	−0.0148	−0.0260	−0.00654	0.0210	0.00259
	(0.233)	(0.0584)	(0.234)	(0.0587)	(0.306)	(0.0378)
Intercept	−0.247		−0.252		−0.932***	
	(0.199)		(0.200)		(0.246)	
Wald χ^2	19.75**		21.03***		13.80**	
Likelihood ratio	23.51***		23.54***		8.93	

Robust standard errors in parentheses
* $p < 0.05$ ** $p < 0.01$ *** $p < 0.001$
[a]FemalePI and FemaleOwner are not used in the model due to lack of variation

Table 7.8 Variance inflation factors

	VIF	
FemalePI	1.002	
FemaleOwner		1.013
Employees	1.044	1.050
SimAwardsDummy	1.049	1.049
ProfInvolved	1.066	1.067
DMT	1.057	1.063

marginal effects of *SimAwardsDummy* on *Failure* are significant at the 0.001 level. The marginal effect of *SimAwardsDummy* in the *FemalePI* specification is −0.242 and when *FemaleOwner* is in the model, the marginal effect is −0.241. These results infer a reduction of 24 percentage points in the probability of *Failure* if at least one similar award was received.

Moving to the model of *TechFailure*, as also shown in Table 7.5, neither *FemalePI* or *FemaleOwner* are included in the estimation due to lack of variation; there are no projects that experienced *TechFailure* and had a female PI or a female owner. The coefficient for *Employees* in the *TechFailure* model is negative, the opposite sign of its estimates in the *Failure* regressions. Although this parameter is not statistically significant, the negative relationship of *Employees* with *TechFailure* suggests that increasing the size of the firm by adding employees reduces the probability of *TechFailure*. The coefficient on *SimAwardsDummy* in the *TechFailure* specification is negative and statistically significant at the 0.05 level. The marginal effect of *SimAwardsDummy* on *TechFailure* of −0.0994 implies a nearly 10 percentage point reduction in the probability of *TechFailure* if any similar award was received as opposed to when zero similar awards were received. The marginal effect of *SimAwardsDummy* in the *Failure* model, discussed above, is more than twice the size of the marginal effect of *SimAwardsDummy* in the *TechFailure* model. Having received any past awards may have a greater impact on *Failure* than on *TechFailure* since *Failure* is a broader measure. The experience gained through the award process and resulting R&D work may provide experiences such as budgeting or project management skills that reduce the likelihood of *Failure* but are not able to reduce the likelihood of *TechFailure*.

7.3.2 Set Two Model Results

The next set of models augments the first set of estimations of Eq. (7.1) by including another project characteristic variable used as an additional measure of human capital, namely *ProfInvolved.* The results from including this additional measure of human capital are provided in Table 7.6. The parameter estimates for both *Failure* models are statistically significant at the 0.05 level and are negative. The marginal effect of *ProfInvolved* in the *FemalePI* specification is −0.171, and similarly in the *FemaleOwner* model, it is −0.170. These results suggest that having university faculty working on the project reduces the probability of *Failure* by 17 percentage points. The marginal effects of *SimAwardsDummy* in both *Failure* models are larger than the *ProfInvolved* marginal effects, implying that receiving at least 1 previous Phase II award translates to a larger reduction in the probability of *Failure* than having university faculty working on the project. The parameter estimate of *ProfInvolved* in the *TechFailure* regression is negative, however, is not statistically significant.

7.3.3 Set Three Model Results

To introduce another measure of project complexity into the analysis, the former set of models are each augmented with *DMT,* a fixed effect control variable that indicates when the technology being developed by each project should be classified into multiple technology sectors. Table 7.7 provides the model results with *DMT* as an additional regressor to the specifications shown in Table 7.6. The parameter estimates for *DMT* in both *Failure* specifications are negative, however, not statistically significant. The hypothesized relationship between *DMT* and *Failure* was positive, based on the notion that a technology that falls in more than one technology sector could be more complex than one classified in a single sector, thus these estimates do not support the original hypothesis.

As shown in Table 7.7, the coefficient on *DMT* in the *TechFailure* model is positive, opposite to that found in the *Failure* specifications, though not statistically significant. This result is in line with the hypothesis that projects classified into multiple technology sectors are more complex on average and, ceteris paribus, have higher failure rates.

7.3.4 Variance Inflation Factors

Finally, variance inflation factors (VIFs) for the group of firm and project characteristics that were used in the prior three sets of estimations of Eq. (7.1) (Tables 7.5, 7.6, 7.7) are shown in Table 7.8. As shown in Table 7.8, the VIF on each variable is only slightly greater than 1; this suggests that there is little concern for multicollinearity between these variables. Since these sets of variables encompass all the variables used in the three sets of estimations of Eq. (7.1), multicollinearity is not a concern for any of the prior estimations of Eq. (7.1).

7.4 Conclusion

This chapter presented the initial empirical analyses to examine firm and project characteristics and their associations with SBIR project failure. From this analysis, it has been shown that relevant experience, as measured by having received a prior SBIR award, has a statistically significant and negative relationship with SBIR project failure. This result is intuitive because having relevant experience should help firms navigate or avoid pitfalls in the R&D process that may lead to project failure by drawing on their relevant experience.

Additionally, this analysis has shown that there is a statistically significant, negative relationship between human capital, as measured by a university professor's involvement in the SBIR project, and project failure. Being a university professor signals a high level of human capital given the educational requirement to be a professor. This result is also intuitive because the additional human capital a university professor provides to the firm may translate to a firm that is more capable, innovative, and productive, which ultimately reduces the likelihood of project failure.

In the next chapter, the initial model results will continue to be expanded upon by introducing covariates with SBIR project failure that are new to the literature. Specifically, characteristics of the principal investigators that were involved with the random sample of SBIR projects used in this analysis will be added to the initial model to enrich the reduced form specification discussed earlier.

Chapter 8
Principal Investigators' Experience in Collaborative R&D: Empirical Findings

Abstract This chapter introduces principal investigators (PIs) as a new-to-the-literature covariate with publicly funded research and development (R&D) project failure. The available demographic data on PIs are described with initial hypotheses of their relationship with failure. The demographic variables available include the age, nationality, and ethnicity of the PI. Using the model formulation discussed earlier, PI demographic data are included in the estimations to augment the firm and project characteristic variables discussed in the previous chapter. Additionally, data on the leadership role(s) the PI has in the firm and their professional background are also described with initial hypotheses of their relationship with failure. These data include being a PI, chief executive officer (CEO), and founder of the firm, as well as information on having an academic or business background. Estimation results show that the age of the PI is the only demographic variable that has a significant relation with R&D project failure. Additionally, the following analysis shows that PIs who hold multiple leadership roles in the firm such as being the PI, CEO, and founder are more likely to experience R&D project failure than other PIs.

8.1 Introduction

The following chapter builds on the analysis performed in Chap. 7 by examining three additional dimensions to measure PI experience that further enrich the firm and project characteristics discussed in Chap. 7. The additional dimensions considered that are specific to PIs are demographic factors, PIs' role as firm leaders and chief executives, and the homophilic relationship between PIs and firm owners. Four types of demographic variables are available in the data: gender, ethnicity, age, and immigration status or nationality. Considering firm leadership characteristics, the data provide information from which it can be determined if the PI was the CEO, a firm founder, or both the CEO and a founder. Additional information about founders is available, such as the number of founders with a business or academic background. Further, as discussed in Chap. 7, the data provide information on the gender

M. Boyce, *Principal Investigators and R&D Failure*, Academic Entrepreneurship, Scientists, and Scientific Careers, https://doi.org/10.1007/978-3-031-43608-6_8

of both the owner and the PI, which is used to analyze the homophilic relationship between firm owners and PIs. Table 8.1 provides definitions of the PI experience variables.

Specifically, in the following chapter, Eq. (7.1) is used once again to estimate the relationship between each measure of SBIR project failure (*Failure* and *TechFailure*) and the firm and project characteristics discussed in Chap. 7, augmented with additional measures of PI experience. Therefore, ***X***, from Eq. (7.1), represents a vector of firm and project characteristics plus measures of the PI's experience as drawn from the random sample. The ***X*** vector used in the reduced form specifications of the structural model (laid out in Chap. 5) contains elements of both ***Q*** (research and development (R&D) inputs) and ***E*** (experience of the PI and/or firm). The models used in this chapter build on those examined in Chap. 7 by providing a richer formulation of ***E*** through additional measures of PI experience.

The chapter is organized as follows: Hypotheses of the relationship between the additional PI variables and project failure are discussed followed by descriptive statistics of the PI demographic variables. Estimation results from augmenting the experience vector from Chap. 7 with PI demographic variables are provided next, including an interpretation of the parameter estimates. Continuing to enrich the experience vector of the PI and firm, PI leadership and vocational background factors are described and then used to augment the prior model specifications. The final analysis provided in this chapter takes an analytical approach to draw insights into the homophilic relationship between PIs and owners of the firm.

Table 8.1 PI Variable definitions

Variable	Definition
FemalePI	= 1 if the principal investigator is a female; 0 otherwise
MinorityPI	= 1 if the principal investigator is a minority; 0 otherwise
Age30DecilePI	= 1 if the principal investigator is between 30 and 39 years old; 0 otherwise
Age40DecilePI	= 1 if the principal investigator is between 40 and 49 years old; 0 otherwise
Age50DecilePI	= 1 if the principal investigator is between 50 and 59 years old; 0 otherwise
AmerPI	= 1 if the principal investigator is an American-born US citizen; 0 otherwise
PICEO	= 1 if the principal investigator is the CEO of the firm; 0 otherwise
PIFounder	= 1 if the principal investigator is a founder of the firm; 0 otherwise
PIFounderCEO	= 1 the principal investigator is a founder and the CEO of the firm; 0 otherwise
PIOneFounder	= 1 if the principal investigator is the only founder of the firm; 0 otherwise
PIOneFounderCEO	= 1 if the principal investigator is the only founder and the CEO of the firm; 0 otherwise
BizBackground	Number of founders with a background in business
AcademicBackground	Number of founders with an academic background

8.1.1 PI Hypothesized Relationships with Project Failure

As discussed previously, gender has been identified in prior studies as having significant explanatory power of project failure. Although the results of the initial probit estimations (Tables 7.5, 7.6, and 7.7) for the variable *FemalePI* are not statistically significant at conventional levels, the negative association with project failure is directionally consistent with prior studies. Therefore, it is expected that the inverse relationship between project failure and *FemalePI* remains unchanged after including additional experience measures.

The relationship between the ethnicity of the PI and project failure has not been established in the literature. The factors that distinguish ethnicities do not intuitively suggest a differing relationship among ethnicities and project failure. Therefore, no assumption is offered in this book about the association between ethnicity and failure.

The relationship between the age of the PI and failure has not been established in the literature. Age, often used as a proxy for experience, does not have a clear linear relationship with project failure. It is possible that older PIs have more experience and through this experience tend to fail less often than their younger counterparts. However, at some point it is possible that being too old, or having too much experience, results in a greater likelihood of project failure. This may come to fruition due to increased risk-taking by older, more experienced PIs who have had significant success in the past. Therefore, no assumption is offered in this analysis about the association between ethnicity and age of the PI.

Similar to the ethnicity of the PI, the immigration status of the PI and its relationship with project failure have not been established in the literature. It is not obvious whether a certain immigration status of the PI should increase the likelihood of project failure compared to another. Hence, no assumption is offered about the association between the immigration status of the PI and project failure.

The relationship between project failure and whether the PI was also the CEO or a firm founder has not been established in the literature. However, Bednar et al. (2019) estimated the probability of commercialization while controlling for the situation when the PI was also the CEO. They found a statistically significant negative relationship between the probability of commercialization and if the PI was also the CEO. This relationship loosely implies a positive relationship between firms with a PI CEO and project failure. Therefore, based on this implication, a positive relationship between *PICEO* and *Failure* is expected. Similarly, it is expected of *PIFounder* to also have a positive relationship with failure. One explanation for this hypothesis is that intuitively, PIs may be more likely to hold multiple leadership roles in smaller firms where resources are less abundant, and the demands of multiple top leadership positions do not allow the PI to succeed as either a scientist or firm founder. Firms where the PI is performing triple duty, that is PI, founder, and CEO, are also hypothesized to have a positive relationship with *Failure* due to the previous expectations of *PICEO* and *PIFounder*. The rationale behind these hypotheses suggests firms with a *PIFounderCEO* should have a greater probability of *Failure* than those with either a *PICEO* or a *PIFounder*. Similarly, firms that were founded solely by the PI,

PIOneFounder, are expected to be positively associated with *Failure* and have a greater probability of *Failure* than those with a *PIFounder*. Finally, firms with a *PIOneFounderCEO* are expected to have the highest probability of *Failure* in accordance with the hypotheses discussed above.

Data are available on the background of the firm founders: either business or academic. Andersen et al. (2017), using similar data, considered the business background of firm owners when estimating the probability of SBIR project failure and found a statistically significant inverse relationship with failure. However, this empirical analysis uses a slightly different measure of vocational experience. Because the focus of this analysis is on the PI's role in project failure and not necessarily a founder or CEO, founders who were also the PI (as well as PI CEOs) are used in the experience vector as opposed to firm founders unconditional on being a PI. Nonetheless, given the closeness in measures between Andersen et al. (2017) and those used in this analysis, a negative relationship between PI founders with a business background and project failure is expected. It is expected that firms with PI founders who have an academic background are inversely related to failure given the significant negative relationship estimated between *Failure* and university faculty involvement in the project (see Table 7.6).

Additionally, Bednar et al. (2019) examined the homophilic relationship between firm owners and PIs and found that female PIs tend to perform better as measured by the probability of commercializing when the firm owner is also female; this association will also be explored in this chapter; however, the measure of concern is project failure. Hence, it is expected that the relationship between homophilic firms (measured by owner and PI) and project failure is directionally consistent with the relationship found by Bednar et al. (2019).

8.2 PI Demographics

8.2.1 Descriptive Statistics

The distributions of PIs conditional on each of the four demographic categories from the random sample of 169 SBIR Phase II–funded projects are presented in Table 8.2. Female PIs comprise 5.3 percent of the sample and minority PIs make up 10.1 percent of the sample. The remaining 85 percent of projects had non-female, non-minority PIs.

Considering a further disaggregation of the minority binary variable, PIs of the Asian-Indian ethnicity represent 7.1 percent of the 169 randomly sampled projects and 70.6 percent of the minority PIs. PIs of the Asian-Pacific ethnicity comprise 2.4 percent of the random sample and 23.5 percent of the minority PIs. The remaining proportion of minority PIs are Hispanic, making up less than 1 percent of the random sample and 5.9 percent of the minority PIs. There are no PIs in the sample that fall into the Black, Native American, or other ethnicities.

Table 8.2 Descriptive statistics of PI demographic variables

Characteristic	Number	Percent of sample ($n = 169$)
Gender/Ethnicity of PI		
Female PI	9	5.3254
Minority PI	17	10.0592
Neither female nor minority PI	144	85.2071
Ethnicity of minority PI		
Asian-Indian	12	7.1006
Asian-Pacific	4	2.3669
Black	0	0
Hispanic	1	0.5917
Native American	0	0
Other	0	0
Age of PI		
< 25	0	0
25–29	1	0.5917
30–34	7	4.1420
35–39	28	16.5680
40–44	32	18.9349
45–49	25	14.7929
50–54	24	14.2012
55–59	31	18.3432
60–64	11	6.5089
65+	10	5.9172
Immigration status of PI		
American-born US citizen	105	62.1302
Naturalized US citizen	34	20.1183
US green card	27	15.9763
H1 visa	3	1.7751
Other	0	0

There are ten age brackets for PIs; the youngest category being less than 25 years of age and the oldest greater than 65. There are no PIs in the sample who were less than 25 years old at the time of the survey, and less than 1 percent of PIs were between 25 and 29 years old. PIs who were between 30 and 34 years old represent 4.1 percent of the random sample, and PIs with ages between 35 and 39 make up 16.6 percent of the sample. PIs between the ages of 40 and 44 are the most frequent in the sample, representing 18.9 percent and PIs between 45 and 49 years old make up 14.8 percent of the sample. The age group 50–54 comprises 14.2 percent of the sample and PIs between 55 and 59 years old comprise 18.3 percent of the random sample, the second most frequent age range of PIs. PIs between the ages of 60 and 64 represent 6.5 percent of the sample and PIs ages 65 and older comprise 5.9 percent of the random sample.

The most frequent immigration status of PIs in the random sample is that of American-born US citizens: these PIs represent 62.1 percent of the sample. PIs who were naturalized US citizens made up 20.1 percent of the sample and those who held a green card comprised 16.0 percent. Finally, PIs who held an H1 visa make up 1.8 percent of the sample, and there were no PIs who had an immigration status other than those mentioned above.

To begin to explore the relationship between PI demographic variables and *Failure*, the random sample of 169 Phase II–funded projects was segmented by those that failed (*n* = 36) and those that did not (*n* = 133). As shown in Table 8.3, less than 1 percent of firms with a female PI experienced *Failure,* the smallest percent of *Failure* among the gender/ethnicity dimension. Further, conditional on a firm having a female PI (*n* = 9), 11 percent experienced *Failure*. Firms with a minority PI (*n* = 17) and that experienced *Failure* comprise 2.96 percent of all firms in the sample, while 29 percent of firms that had a minority PI experienced *Failure.* Firms with a non-female, non-minority PI (*n* = 144) and experienced *Failure* comprise 17.6 percent of the random sample and experienced a 21 percent *Failure* rate within the subsample of firms that had a non-female, non-minority PI.

Moving to the age groups of the PI, there is not a firm within the random sample that had a PI in the age range of 30–34 (*n* = 7) and that experienced *Failure,* though PIs in this age bracket account for about 4 percent of the total random sample. Firms with a PI in the 35–39 age range (*n* = 28) and that encountered *Failure* comprise nearly 3 percent of the sample though almost 18 percent of PIs within the 35–39 age bracket experienced *Failure.* Firms with a PI that was between 40 and 44 years old (*n* = 32) and that met *Failure* account for 5.3 percent of the total sample, the largest share of *Failure* by age indicator. Further, 28 percent of PIs 40–44 years old experienced *Failure*, which is the third highest failure rate within an age cohort. Firms with a PI age range 45–49 (*n* = 25) and that encountered *Failure* comprise 4.7 percent of all firms in the sample, the second largest share of firms that failed conditional on the age of the PI. The *Failure* rate of 32 percent among PIs 45–49 years old is the second highest rate across all age cohorts. Firms with a PI between 50 and 54 years old (*n* = 24) and that experienced *Failure* represent 1.2 percent of the total sample. Within this age range, less than 1 percent of PIs failed, which is the second lowest *Failure* rate within an age cohort. Firms with a PI between 55 and 59 years old (*n* = 31) and that experienced *Failure* make up almost 3 percent of the random sample, though conditional on firms with a PI in this age bracket, 16 percent of them failed. The second oldest age cohort of PIs are those between 60 and 64 years old (n = 11). PIs in this cohort that encountered *Failure*, were PIs for 2.4 percent of all firms in the sample and experienced the highest failure rate within an age cohort. Specifically, 36 percent of PIs ages 60–64 experienced *Failure.* The oldest cohort of PIs ages 65 years and older (*n* = 10) and that witnessed *Failure* represent 1.2 percent of all PIs in the sample, which translates to a 20 percent *Failure* rate among the oldest PIs in the sample.

The last demographic variable considered in this analysis is the immigration status of the PI. Again, as shown in Table 8.3, PIs who are American-born US citizens (*n* = 105) and whose firms experienced *Failure* represent 10.6 percent of the

Table 8.3 Distribution of PI demographics by *Failure*

	Failure					
	1	0	1	0	1/(0,1)	Count
Characteristic	Percent (n = 169)		Rate	**Gender/ Ethnicity of PI**	Female PI	1
8	0.6%	4.7%	11.1%	Minority PI	5	12
3.0%	7.1%	29.4%	Neither female nor minority PI	30	114	17.8%
67.5%	20.8%	**Ethnicity of minority PI**	Asian-Indian	2	10	1.2%
5.9%	16.7%	Asian-Pacific	3	1	1.8%	0.6%
75.0%	Black	0	0	0.0%	0.0%	0.0%
Hispanic	0	1	0.0%	0.6%	0.0%	Native American
0	0	0.0%	0.0%	0.0%	Other	0
0	0.0%	0.0%	0.0%	**Age of PI**	< 25	0
0	0.0%	0.0%	0.0%	25–29	1	0
0.6%	0.0%	100.0%	30–34	0	7	0.0%
4.1%	0.0%	35–39	5	23	3.0%	13.6%
17.9%	40–44	9	23	5.3%	13.6%	28.1%
45–49	8	17	4.7%	10.1%	32.0%	50–54
2	22	1.2%	13.0%	8.3%	55–59	5
26	3.0%	15.4%	16.1%	60–64	4	7

(continued)

Table 8.3 (continued)

Characteristic	*Failure*					
	1	0	1	0	1/(0,1)	Count
	Percent ($n = 169$)		Rate	**Gender/ Ethnicity of PI**	Female PI	1
2.4%	4.1%	36.4%	65+	2	8	1.2%
4.7%	20.0%	**Immigration status of PI**	American-born US citizen	18	87	10.7%
51.5%	17.1%	Naturalized US citizen	9	25	5.3%	14.8%
26.5%	US green card	9	18	5.3%	10.7%	33.3%
H1 visa	0	3	0.0%	1.8%	0.0%	Other
0	0	0.0%	0.0%	0.0%		

total sample, the largest share of PIs conditional on immigration status. Within the American-born US citizen PIs, 17 percent of their projects failed. PIs who are naturalized US citizens ($n = 34$) and who experienced *Failure* account for 5.3 percent of all PIs in the sample and have a *Failure* rate among this immigration cohort of 26.5 percent. PIs who held a US green card ($n = 27$) and encountered *Failure* represent 5.3 percent of PIs in the full sample, while they have a within-cohort *Failure* rate of 33.3 percent. Finally, PIs who held an H1 visa ($n = 3$) did not have any cases of *Failure*.

To consider the relationship among PI demographic variables and *TechFailure,* the random sample of 169 Phase II–funded projects was also segmented by those that failed for technical reasons ($n = 12$) and those that did not ($n = 157$). As shown in Table 8.4, no firms with a female PI experienced *TechFailure*. Firms with a minority PI ($n = 17$) and that experienced *TechFailure* represent 1.8 percent of the random sample, while 17 percent of minority PIs failed for technical reasons. Firms with a non-female, non-minority PI ($n = 144$) and that experienced *TechFailure* comprise 5.3 percent of the random sample and had a *TechFailure* rate of 6.5 percent among non-female, non-minority PIs.

Moving to the age bins of the PI, no firm had a PI less than 35 years old and experienced *TechFailure*. Firms that had a PI in any one of the age buckets 35–39, 40–44, or 45–49 and that experienced *TechFailure* each comprised 1.8 percent of the full sample. In terms of *TechFailure* prevalence, 10.7 percent of 35–39-year-old PIs, 9.4 of the 40–44-year-old PIs, and 12 percent of the 45–49-year-old PIs failed for technical reasons. There were no firms with a PI between 50 and 54 years old and that experienced *TechFailure*. Firms with a PI between 55 and 59 years old and that experienced *TechFailure* make up less than 1 percent of the full sample, though conditional on firms with a PI in this age bracket, 3.2 percent of them failed for technical reasons. PIs between 60 and 64 years old and that encountered *TechFailure* make up 1.2 percent of all PIs, which translates to an 18.2 percent *TechFailure* rate among this age cohort. The oldest cohort of PIs ages 65 years and older did not have any cases of failure for technical reasons.

Finally, considering the immigration status of the PI segmented by *TechFailure*, as shown in Table 8.4, PIs who are American-born US citizens and whose firms experienced *TechFailure* represent almost 3 percent of the full sample. Within the American-born US citizen PIs, 5 percent of their projects failed for technical reasons. PIs who are naturalized US citizens and who experienced *TechFailure* account for 1.2 percent of all PIs in the sample and have a *TechFailure* rate among this immigration cohort of 5.9 percent. PIs who held a US green card and encountered *TechFailure* represent almost 3 percent of PIs in the full sample, while they have a within-cohort *TechFailure* rate of 18.5 percent. Finally, PIs who held an H1 visa did not have any cases of *TechFailure*.

Table 8.4 Distribution of PI demographics by *TechFailure*

	TechFailure				
	1	0	1	0	1/(0, 1)
Characteristic	Count		Percent (n = 169)		Rate
Gender/Ethnicity of PI					
Female PI	0	9	0.0%	5.3%	0.0%
Minority PI	3	14	1.8%	8.3%	17.6%
Neither female nor minority PI	9	135	5.3%	79.9%	6.3%
Ethnicity of minority PI					
Asian-Indian	1	11	0.6%	6.5%	8.3%
Asian-Pacific	2	2	1.2%	1.2%	50.0%
Black	0	0	0.0%	0.0%	0.0%
Hispanic	0	1	0.0%	0.6%	0.0%
Native American	0	1	0.0%	0.6%	0.0%
Other	0	0	0.0%	0.0%	0.0%
Age of PI					
< 25	0	0	0.0%	0.0%	0.0%
25–29	0	1	0.0%	0.6%	0.0%
30–34	0	7	0.0%	4.1%	0.0%
35–39	3	25	1.8%	14.8%	10.7%
40–44	3	29	1.8%	17.2%	9.4%
45–49	3	22	1.8%	13.0%	12.0%
50–54	0	24	0.0%	14.2%	0.0%
55–59	1	30	0.6%	17.8%	3.2%
60–64	2	9	1.2%	5.3%	18.2%
65+	0	10	0.0%	5.9%	0.0%
Immigration status of PI					
American-born US citizen	5	100	3.0%	59.2%	4.8%
Naturalized US citizen	2	32	1.2%	18.9%	5.9%
US green card	5	22	3.0%	13.0%	18.5%
H1 visa	0	3	0.0%	1.8%	0.0%
Other	0	0	0.0%	0.0%	0.0%

8.2.2 Model Results with PI Demographics

Leveraging the reduced form model from Chap. 7, Eq. (7.1), the following estimations include additional factors of PI experience. In terms of the structural form of this model, as presented in Chap. 5, the additional variables represent a richer experience vector, $\boldsymbol{E}$, through the addition of demographic-type information. The additional information on PIs' experience provides more evidence in understanding how variations in PIs' experience are associated with project failure. Further, the focus of the following discussion is on demographic variables of PIs, therefore discussion of estimation results for other variables will be limited unless there is a meaningful change in a variable's output from previous results.

The first additional PI variable examined is the minority status of the PI. This variable is used to augment the model results presented in Table 7.7. Results from the estimation for both *Failure* and *TechFailure* measures including the *MinorityPI* measure are provided in Table 8.5. The estimated parameters for *MinorityPI* in both the *Failure* and *TechFailure* regressions are not statistically significant nor are the estimated marginal effects. These results are in line with the hypothesis that variation between PIs based on ethnicity does not have a significant relationship with SBIR project failure.

Continuing to build on the experience vector, model results from the inclusion of age indicators of the PI are provided in Table 8.6. As shown in Table 8.2, the survey data used in this empirical work provide age indicators in approximately five-year intervals. To use in the model, the age brackets were collapsed into decile bins so that there was a meaningful mass of observations within each bin. The reference age bucket represents PIs who were age 60 and greater. As shown in Table 8.6, the three age indicators all have a negative parameter in the *Failure* estimation, indicating that they are less likely to experience *Failure* relative to PIs at least 60 years old (the oldest cohort); however, only *Age30DecilePI* is statistically significant at conventional levels (0.05 level). *Age50DecilePI* has a p value of 0.052, which is marginally outside the threshold of reported significance in this book, however, is suggestive of a significant relationship. Further, none of the age parameters are statistically significant in the *TechFailure* specification.

Table 8.5 PI Probit with *MinorityPI*

	(1)		(2)	
	Failure	Marginal effects	*TechFailure*	Marginal effects
FemalePI	−0.700	−0.137	–[a]	–[a]
	(0.578)	(0.0830)		
MinorityPI	0.559	0.158	0.829	0.143
	(0.357)	(0.107)	(0.429)	(0.0926)
Employees	0.00219	0.000545	−0.00407	−0.000484
	(0.00236)	(0.000575)	(0.00255)	(0.000354)
SimAwardsDummy	−0.903***	−0.246***	−0.832*	−0.102*
	(0.232)	(0.0631)	(0.333)	(0.0409)
ProfInvolved	−0.641*	−0.144*	−0.476	−0.0471
	(0.320)	(0.0604)	(0.470)	(0.0364)
DMT	−0.0592	−0.0147	0.0171	0.00203
	(0.235)	(0.0581)	(0.312)	(0.0372)
Intercept	−0.272		−1.008***	
	(0.200)		(0.261)	
Wald χ^2	20.29**		18.54**	
Likelihood ratio	25.57***		12.35*	

Robust standard errors in parentheses
*$p < 0.05$ ** $p < 0.01$ *** $p < 0.001$
[a]*FemalePI* is not used in the model due to lack of variation

Table 8.6 PI probit with age decile indicators

	(1)		(2)	
	Failure	Marginal effects	*TechFailure*	Marginal effects
FemalePI	−0.856	−0.149*	–[a]	–[a]
	(0.616)	(0.0746)		
MinorityPI	0.671	0.177	0.714	0.112
	(0.378)	(0.108)	(0.428)	(0.0812)
Age30DecilePI	−0.948*	−0.179**	−0.201	−0.0217
	(0.422)	(0.0632)	(0.500)	(0.0506)
Age40DecilePI	−0.0603	−0.0137	0.0391	0.00450
	(0.374)	(0.0846)	(0.456)	(0.0527)
Age50DecilePI	−0.756	−0.161*	−0.751	−0.0697
	(0.390)	(0.0756)	(0.578)	(0.0427)
Employees	0.00186	0.000427	−0.00512	−0.000586
	(0.00234)	(0.000526)	(0.00281)	(0.000393)
SimAwardsDummy	−1.027***	−0.258***	−0.785*	−0.0931*
	(0.245)	(0.0623)	(0.330)	(0.0372)
ProfInvolved	−0.749*	−0.172*	−0.544	−0.0622
	(0.340)	(0.0740)	(0.449)	(0.0512)
DMT	0.0689	0.0158	0.0810	0.00932
	(0.256)	(0.0586)	(0.304)	(0.0352)
Intercept	0.165		−0.800	
	(0.342)		(0.458)	
Wald χ^2	33.38***		18.65*	
Likelihood ratio	35.94***		15.64*	

Robust standard errors in parentheses
*$p < 0.05$ ** $p < 0.01$ *** $p < 0.001$
[a]*FemalePI* is not used in the model due to lack of variation

The *Age30DecilePI* marginal effect of −0.179 in the *Failure* specification is significant at the 0.01-level and suggests that PIs who are in their 30s are expected to experience a probability of *Failure* 17.9 percentage points less relative to the oldest cohort. Similarly, PIs who fall into *Age50DecilePI* have a probability of *Failure* that is 16.1 percentage points less relative to PIs who are at least 60 years old as shown by the estimated marginal effect. The marginal effect of *Age50DecilePI* is significant at the 0.05 level.

Age is a traditional proxy for experience since often one acquires experience over time. Therefore, it is straightforward that younger PIs are likely less experienced than older PIs as they have had less time to have accumulated human capital. With less experience, the typical projects that younger PIs work on may be inherently less risky than those more experienced PIs would be involved with. Further, younger PIs with less experience should have a higher marginal rate of human capital accumulation than their older cohorts due to diminishing marginal returns on experience. The higher marginal rate of human capital accumulation of younger PIs may translate to a form of motivation in the PI that results in higher productivity.

Simonton (1988) and Simonton (1991) found that scientific productivity peaks at an age just shy of 40 years old and then begins to gradually decline. The decline in productivity may also coincide with human capital depreciation in the oldest cohort of PIs. While the oldest cohort of PIs may have achieved success in their earlier years, being toward the end of their career could lessen the motivation to maintain their skills up to the level required for continued success. Additionally, the projects the most experienced cohort of PIs are typically involved with may be riskier than those of younger PIs. Thus, in summary, younger PIs may have higher levels of motivation while working on less risky projects than the oldest cohort, which helps explains why the younger cohorts of PIs with less experience have lower probabilities of *Failure* relative to the oldest cohort.

The age decile indicator parameters are not statistically significant in the *TechFailure* regression, nor are their marginal effects. Given the nature of *TechFailure,* it is reasonable that the typical experience gained that is attributable to the age of the PI does not necessarily translate to experience that helps reduce the likelihood of *TechFailure*.

Moving forward, the last demographic variable considered here to augment the experience vector is the nationality of the PI. Table 8.7 presents the model results with *AmerPI* as an additional regressor to those presented in Table 8.6. As expected, the parameter estimate for the nationality of the PI is not statistically significant nor is the marginal effect in either the *Failure* or the *TechFailure* specifications. Although Del et al. (2017) found that a country's culture can influence a PI to have a greater preference toward entrepreneurial activities, the country a PI comes from does not necessarily equip a PI with the experience needed to reduce the probability of failure.

8.3 PIs as CEOs/Firm Founders

8.3.1 Descriptive Statistics

Summary statistics for PIs who are also firm leaders are provided in Table 8.8. As shown, PIs who are also the CEO, *PICEO,* represent 22.5 percent of the random sample of Phase II–funded SBIR projects used in this analysis. PIs who were also a founder, and not necessarily the only founder, make up 29.6 percent of the sample. PIs who were the CEO and a firm founder, *PIFounderCEO*, comprise 21.3 percent of the sample. Summary statistics are also provided for a subsample of *PIFounder,* specifically, PIs who were the sole founder of the firm, *PIOneFounder*. PIs who were the sole firm founder represent 13.6 percent of the sample. Similarly, PIs who were the CEO and the sole firm founder, *PIOneFounderCEO,* represent 11.2 percent of the sample.

As mentioned previously, the second-round survey also provides information on firm founders' background, specifically the number of founders with a business or

Table 8.7 PI probit with nationality indicator

	(1)		(2)	
	Failure	Marginal effects	*TechFailure*	Marginal effects
FemalePI	−0.855	−0.147	–[a]	–[a]
	(0.651)	(0.0779)		
MinorityPI	0.428	0.106	0.538	0.0777
	(0.396)	(0.106)	(0.436)	(0.0753)
Age30DecilePI	−1.077*	−0.195**	−0.296	−0.0311
	(0.427)	(0.0595)	(0.479)	(0.0461)
Age40DecilePI	−0.0867	−0.0194	0.00284	0.000324
	(0.382)	(0.0846)	(0.457)	(0.0521)
Age50DecilePI	−0.824*	−0.171*	−0.817	−0.0749
	(0.398)	(0.0740)	(0.605)	(0.0434)
AmerPI	−0.421	−0.0991	−0.305	−0.0359
	(0.263)	(0.0631)	(0.304)	(0.0356)
Employees	0.00143	0.000323	−0.00586*	−0.000669
	(0.00233)	(0.000519)	(0.00291)	(0.000405)
SimAwardsDummy	−0.964***	−0.238***	−0.741*	−0.0866*
	(0.242)	(0.0587)	(0.321)	(0.0350)
ProfInvolved	−0.753*	−0.170*	−0.537	−0.0612
	(0.339)	(0.0727)	(0.454)	(0.0514)
DMT	0.0466	0.0105	0.0599	0.00687
	(0.255)	(0.0576)	(0.301)	(0.0346)
Intercept	0.488		−0.544	
	(0.414)		(0.493)	
Wald χ^2	30.91***		16.36	
Likelihood ratio	38.22***		16.27	

Robust standard errors in parentheses
$^*p < 0.05$ $^{**}p < 0.01$ $^{***}p < 0.001$
[a]*FemalePI* is not used in the model due to lack of variation

Table 8.8 Descriptive statistics of PIs as firm leaders

Indicator	Mean	StdDev	Range
PICEO	0.2249	0.4187	0/1
PIFounder	0.2959	0.4578	0/1
PICEOFounder	0.213	0.4107	0/1
PIOneFounder	0.1361	0.3439	0/1
PIOneFounderCEO	0.1124	0.3168	0/1
BizBackground	0.7574	1.0552	0–5
AcademicBackground	1.0592	1.3615	0–10

academic background. As shown in Table 8.8, the mean number of firm founders with a background in business is slightly less than 1, at 0.7574. The standard deviation is slightly greater than 1 founder, and within the sample, the number of founders with a business background ranged between 0 and 5. The mean number of founders with an academic background is slightly more than 1 at 1.0592 founders, with a standard deviation of 1.3615. The number of founders with an academic background ranged between 0 and 10 founders within the random sample of firms.

To begin exploring the relationship between the firm leadership experience variables and *Failure,* descriptive statistics of the firm leadership variables are segmented by those that experienced project failure (*Failure* = 1) and those that did not (*Failure* = 0). As shown in Table 8.9, 5.9 percent of the sample were a *PICEO* that experienced *Failure,* while *PICEO* that did not fail made up 16.6 percent of the sample. Therefore, firms with a *PICEO* had a failure rate of 26.3 percent. Firms with a *PIFounder* and that experienced *Failure* comprise 5.9 percent of the sample, while their counterparts that did not experience *Failure* comprise 23.7 percent of the firms sampled. Firms with a *PIFounder* have a *Failure* rate of 20.0 percent, the lowest rate of failure among the firm leadership experience variables. Firms with a *PIFounderCEO* and that experienced *Failure* also make up 5.9 percent of the sample, although *PIFounderCEO* firms that did not fail hold 15.4 percent of the sample. This translates to a 27.8 percent *Failure* rate for *PIFounderCEO* firms. *PIOneFounder* firms that had *Failure* comprise 4.7 percent of the random sample and those that did not comprise 8.9 percent. Thus, *PIOneFounder* firms had a *Failure* rate of 34.8 percent, which is 14.8 percentage points greater than *PIFounder* firms. Firms that were founded and managed by the PI, *PIOneFounderCEO,* and that experienced *Failure* comprise 4.7 percent of the sample and those that did not fail comprise 6.5 percent, which translates to the highest *Failure* rate among the various firm leadership experience types of 42.1 percent. These statistics are suggestive of the hypothesis that the more concentrated the leadership experience is within a firm, the more likely they are to experience *Failure.*

The mean number of firm founders with a business background, *BizBackground,* conditional on having experienced *Failure* is 0.6389, which is less than the mean *BizBackground* of the firms that did not fail of 0.7895. This is consistent with the

Table 8.9 Descriptive statistics of PIs as firm leaders grouped by *Failure (standard deviations in parenthesis)*

	Failure				
	1	0	1	0	1/ (0, 1)
Indicator	Count		Percent (*n* = 169)		Rate
PICEO	10	28	5.9%	16.6%	26.3%
PIFounder	10	40	5.9%	23.7%	20.0%
PIFounderCEO	10	26	5.9%	15.4%	27.8%
PIOneFounder	8	15	4.7%	8.9%	34.8%
PIOneFounderCEO	8	11	4.7%	6.5%	42.1%
			***n* = 36**	***n* = 133**	
BizBackground[a]	–	–	0.6389 (0.8669)	0.7895 (1.1013)	–
AcademicBackground[a]	–	–	0.8611 (1.3342)	1.1128 (1.3688)	–

[a]values are means of the variable

hypothesis that *BizBackground* is negatively associated with *Failure*. Similarly, the mean *AcademicBackground* for firms that failed is 0.8611, which is less than those that did not fail of 1.1128. This also is consistent with the hypothesis that *AcademicBackground* is negatively associated with *Failure.*

To begin exploring the relationship between the firm leadership experience variables and *TechFailure,* descriptive statistics of the firm leadership variables are also segmented by those that experienced project failure for technical reasons (*TechFailure* = 1) and those that did not (*TechFailure* = 0). As shown in Table 8.10, 1.2 percent of the sample were a *PICEO* that experienced *TechFailure*, while *PICEO* firms that did not fail made up 21.3 percent of the sample. Therefore, firms with a *PICEO* had a failure rate because of technical reasons of 5.3 percent. Firms with a *PIFounder* and that experienced *TechFailure* comprise 1.2 percent of the sample, while their counterparts that did not experience *TechFailure* comprise 28.4 percent of the firms sampled. Firms with a *PIFounder* have a *TechFailure* rate of 4.0 percent, the lowest rate of *TechFailure* among the firm leadership experience variables. Firms with a *PIFounderCEO* and that experienced *TechFailure* also make up 1.2 percent of the sample, while *PIFounderCEO* firms that did not fail hold 20.1 percent of the sample. This translates to a 5.6 percent *TechFailure* rate for *PIFounderCEO* firms. *PIOneFounder* firms that had *TechFailure* comprise 1.2 percent of the random sample and those that did not comprised 12.4 percent. Thus, *PIOneFounder* firms had a *TechFailure* rate of 8.7 percent, which is more than twice the rate of firms with a *PIFounder*. Firms that were founded and managed by the PI, *PIOneFounderCEO,* and that experienced *TechFailure* also comprise 1.2 percent of the sample and those that did not fail comprise 10.1 percent, which translates to the highest *TechFailure* rate among the various firm leadership experience types of 10.5 percent. These results are consistent with the hypothesis: less variability of the human capital in key leadership roles for the firm results in a greater likelihood that the firm will experience project failure.

Table 8.10 Descriptive statistics of PIs as firm leaders grouped by *TechFailure (standard deviations in parenthesis)*

	TechFailure				
	1	0	1	0	1/(0, 1)
Indicator	Count		Percent (n = 169)		Rate
PICEO	2	36	1.2%	21.3%	5.3%
PIFounder	2	48	1.2%	28.4%	4.0%
PICEOFounder	2	34	1.2%	20.1%	5.6%
PIOneFounder	2	21	1.2%	12.4%	8.7%
PIOneFounderCEO	2	17	1.2%	10.1%	10.5%
			n = 12	**n = 157**	
BizBackground[a]	–	–	0.75 (0.866)	0.758 (1.0706)	–
AcademicBackground[a]	–	–	0.6667 (0.8876)	1.0892 (1.3885)	–

[a]Values are means of the variable

The mean number of firm founders with a business background conditional on having experienced *TechFailure* is 0.75, which is marginally less than the mean *BizBackground* of the firms that did not fail of 0.758. Given the lack of variation between these two means, it is not clear whether *BizBackground* has a meaningful association with *TechFailure*. The mean *AcademicBackground* for firms that failed for technical reasons is 0.6667, which is less than those that did not fail of 1.0892. This also is consistent with the hypothesis that *AcademicBackground* is negatively associated with *TechFailure*.

8.3.2 Model Results with PI as Firm Leaders

The following set of models continues to build on the reduced form specification discussed previously by augmenting the model with the PI firm leadership measures of experience. In terms of the structural formulation discussed in Chap. 5, adding the information about the leadership of the firm represents experience of both the firm and the PI. This final set of models that incorporates the firm's leadership experience provides the richest experience vector to understand covariates of project failure. Further, the focus of the following discussion is on the firm leadership variables, therefore discussion of estimation results for other variables will be limited unless there is a meaningful change in a variable's output from previous results.

Table 8.11 presents the model results of three estimations of *Failure* with additional regressors to those presented in Table 8.7. The first specification includes *PIFounder*, which has a positive coefficient though the parameter, and the marginal effect is not statistically significant at conventional levels.

The second model results shown in Table 8.11 adds *BizBackground* to the estimation. The parameter of *BizBackground* is not statistically significant, but the estimate is directionally consistent with the hypothesis of a negative relationship with *Failure*. The marginal effect of *BizBackground* is not statistically significant at the reported levels with a *p* value of 0.088, which is outside the bounds of statistical significance used in this analysis. An interaction term between *PIFounder* and *BizBackground* was also included. The coefficient on the interaction term has a *p* value of 0.051, marginally outside the reported significance levels, and has a negative relationship with *Failure*. The interaction term does not provide a direct link to the background of the PI since there may be cofounders; however, the term still provides useful information. It suggests that when a firm has a *PIFounder,* having founders with a background in business at least partially offsets the positive effect on *Failure* from having a *PIFounder*.

The third model results shown in Table 8.11 include *AcademicBackground* and its interaction with *PIFounder*; however, neither the interaction term nor *AcademicBackground* parameter estimates are statistically significant. The marginal effect of *AcademicBackground* has a *p* value of 0.064, which is marginally outside the bounds of the reported significance levels in this analysis, though may be suggestive of a significant relationship. Nonetheless, the marginal effect of

Table 8.11 PI as founder *Failure* probit

	(1)		(2)		(3)	
	Failure	Marginal effects	*Failure*	Marginal effects	*Failure*	Marginal effects
PIFounder	0.112	0.0256	0.394	0.00950	0.425	0.0165
	(0.295)	(0.0679)	(0.367)	(0.0653)	(0.361)	(0.0663)
BizBackground			−0.0443	−0.0506		
			(0.137)	(0.0297)		
PIFounder × *Biz Background*			−0.692			
			(0.355)			
AcademicBackground					−0.107	−0.0501
					(0.149)	(0.0271)
PIFounder × *Academic Background*					−0.433	
					(0.271)	
FemalePI	−0.857	−0.148	−0.707	−0.125	−1.014	−0.161*
	(0.658)	(0.0785)	(0.653)	(0.0877)	(0.733)	(0.0746)
MinorityPI	0.426	0.106	0.384	0.0904	0.505	0.122
	(0.392)	(0.105)	(0.388)	(0.0974)	(0.418)	(0.108)
Age30DecilePI	−1.069*	−0.194**	−1.157**	−0.203***	−1.159**	−0.202***
	(0.430)	(0.0602)	(0.438)	(0.0586)	(0.423)	(0.0561)
Age40DecilePI	−0.0831	−0.0186	−0.114	−0.0245	−0.113	−0.0243
	(0.384)	(0.0850)	(0.401)	(0.0850)	(0.370)	(0.0786)
Age50DecilePI	−0.838*	−0.174*	−0.924*	−0.181**	−0.989*	−0.195**
	(0.393)	(0.0728)	(0.410)	(0.0701)	(0.389)	(0.0682)
AmerPI	−0.417	−0.0982	−0.420	−0.0949	−0.368	−0.0831
	(0.263)	(0.0626)	(0.268)	(0.0612)	(0.263)	(0.0602)
Employees	0.00170	0.000386	0.00203	0.000441	0.00229	0.000499
	(0.00246)	(0.000546)	(0.00246)	(0.000525)	(0.00264)	(0.000563)
SimAwardsDummy	−0.980***	−0.241***	−0.956***	−0.225***	−0.991***	−0.234***
	(0.247)	(0.0594)	(0.252)	(0.0581)	(0.255)	(0.0579)
ProfInvolved	−0.765*	−0.173*	−0.748*	−0.163*	−0.773*	−0.169*
	(0.348)	(0.0744)	(0.356)	(0.0728)	(0.357)	(0.0730)
DMT	0.0731	0.0165	0.0800	0.0174	0.105	0.0228
	(0.258)	(0.0581)	(0.270)	(0.0585)	(0.265)	(0.0575)
Intercept	0.440		0.492		0.545	
	(0.430)		(0.452)		(0.423)	
Wald χ^2	30.42**		38.35***		36.31***	
Likelihood ratio	38.37***		42.81***		43.23***	

Robust standard errors in parentheses
*$p < 0.05$ ** $p < 0.01$ *** $p < 0.001$

AcademicBackground on *Failure* of −0.0501 suggests a 5.0 percentage point reduction in the probability of *Failure* for each additional firm founder that had an academic background, ceteris paribus.

The next two model specifications consider a subset of the *PIFounder* variable using a dichotomous indicator, *PIOneFounder,* to indicate when the PI was the sole founder of the firm. Model results from the inclusion of *PIOneFounder* are shown in Table 8.12. The parameter estimate for *PIOneFounder* is positive as expected, and though not statistically significant at the reported levels, it has a p value of 0.058. The marginal effect of *PIOneFounder* has a p value of 0.080, again outside the thresholds of reported statistical significance in this analysis. Although the statistical significance of the marginal effect of *PIOneFounder* is outside the threshold

Table 8.12 PI as single founder *Failure* probit

	(1)		(2)	
	Failure	Marginal effects	*Failure*	Marginal effects
PIOneFounder	0.700	0.174	0.461	0.208
	(0.369)	(0.0997)	(0.489)	(0.154)
AcademicBackground			−0.177	−0.0244
			(0.139)	(0.0369)
PIOneFounder × *AcademicBackground*			0.354	
			(0.679)	
FemalePI	−0.713	−0.126	−0.827	−0.139
	(0.653)	(0.0870)	(0.708)	(0.0844)
MinorityPI	0.434	0.104	0.574	0.139
	(0.386)	(0.0994)	(0.397)	(0.102)
Age30DecilePI	−1.087*	−0.192**	−1.281**	−0.215***
	(0.437)	(0.0598)	(0.418)	(0.0521)
Age40DecilePI	−0.0816	−0.0177	−0.192	−0.0407
	(0.395)	(0.0849)	(0.371)	(0.0772)
Age50DecilePI	−0.963*	−0.188**	−1.103**	−0.209**
	(0.412)	(0.0702)	(0.396)	(0.0646)
AmerPI	−0.390	−0.0884	−0.352	−0.0787
	(0.267)	(0.0612)	(0.266)	(0.0599)
Employees	0.00251	0.000550	0.00304	0.000659
	(0.00231)	(0.000492)	(0.00255)	(0.000534)
SimAwardsDummy	−1.026***	−0.243***	−1.033***	−0.240***
	(0.252)	(0.0574)	(0.261)	(0.0565)
ProfInvolved	−0.731*	−0.160*	−0.716*	−0.155*
	(0.345)	(0.0709)	(0.361)	(0.0729)
DMT	0.115	0.0251	0.118	0.0256
	(0.265)	(0.0576)	(0.265)	(0.0571)
Intercept	0.324		0.569	
	(0.413)		(0.407)	
Wald χ^2	30.13**		28.73**	
Likelihood ratio	42.40***		44.77***	

Robust standard errors in parentheses
$*p < 0.05$ $**p < 0.01$ $***p < 0.001$

for reporting; the effect is relatively large. Firms in which the PI is the sole founder have a 17.4 percentage points increase in the probability of *Failure* compared to firms that do not have a *PIOneFounder*.

The second model shown in Table 8.12 includes *AcademicBackground* as an additional measure as well as the interaction between *PIOneFounder* and *AcademicBackground.* Because *PIOneFounder* indicates the sole founder was also the PI, the interaction term provides a direct link to the PI's background. However, the parameter estimates for *PIOneFounder*, *AcademicBackground,* and the interaction term between the two are not statistically significant.

In Table 8.13, model results are shown for four separate estimations that consider the case where the PI was also the CEO. Also included in this set of models is the case where the PI was the CEO and was at least one of the firm founders. The first model results shown in Table 8.13 include *PICEO*, which has a positive parameter estimate, but it is not statistically significant at conventional levels with a *p* value of 0.082. The marginal effect of *PICEO* on *Failure* is not statistically significant.

The second estimation results in Table 8.13 include *PIFounderCEO* as an additional measure of experience. The estimate for *PIFounderCEO* suggests a positive association with *Failure* and parameter estimate is significant at the 0.05 level. The marginal effect of 0.151 is not statistically significant at conventional levels, though it has a *p* value of 0.056. The marginal effect of *PIFounderCEO* suggests firms with a *PIFounderCEO* have a 15.1 percentage point increase in the probability of *Failure* over firms that do not.

The third model results shown in Table 8.13 augment the second model with the inclusion of *BizBackground* and its interaction with *PIFounderCEO*. The coefficient on *PIFounderCEO* is positive, as expected, and statistically significant at the 0.05 level. The coefficients on *BizBackground* and the interaction of *PIFounderCEO* and *BizBackground* are both negative, and the *p* value on the interaction term is 0.055, which is suggestive of a significant relationship with *Failure*. However, the marginal effect of *BizBackground* on *Failure* is not statistically significant at reported levels. These results indicate that firms with a less diverse leadership structure, as measured by the multiple roles of a single person, will have a positive impact on *Failure*. Further, the experience gained from having an additional firm founder with a business background can offset (not necessarily completely) the positive impact on *Failure* from a concentrated leadership structure.

The last model results shown in Table 8.13 include *AcademicBackground* as opposed to *BizBackground* as in the prior estimation discussed. The coefficient on *PIFounderCEO* is positive and significant at the 0.05 level. The coefficients for *AcademicBackground* and the interaction term are not statistically significant.

The final firm leadership experience variable used within a model in this analysis is *PIOneFounderCEO*. Firms with a PI as the sole founder and also the CEO represent the most concentrated firm leadership experience available within these data. As shown in the first model results in Table 8.14, the parameter estimate for *PIOneFounderCEO* is significant at the 0.05 level and has a positive relationship with *Failure*. The marginal effect of *PIOneFounderCEO* on *Failure* is also significant at the 0.05 level and is relatively large at 0.261. This indicates that firms with a

Table 8.13 PI as CEO and PI as founder and CEO *Failure* probit

	(1)		(2)		(3)		(4)	
	Failure	Marginal effects	*Failure*	Marginal effects	*Failure*	Marginal effects	*Failure*	Marginal effects
PICEO	0.540	0.128						
	(0.310)	(0.0757)						
PIFounderCEO			0.629*	0.151	0.945*	0.123	0.848*	0.126
			(0.320)	(0.0789)	(0.397)	(0.0729)	(0.373)	(0.0788)
BizBackground					−0.0425	−0.0459		
					(0.136)	(0.0286)		
PIFounderCEO × *BizBackground*					−0.751			
					(0.392)			
AcademicBackground							−0.141	−0.0479
							(0.149)	(0.0273)
PIFounderCEO × *AcademicBackground*							−0.340	
							(0.330)	
FemalePI	−0.856	−0.146	−0.861	−0.146	−0.718	−0.125	−1.026	−0.160*
	(0.694)	(0.0822)	(0.705)	(0.0828)	(0.674)	(0.0890)	(0.799)	(0.0789)
MinorityPI	0.423	0.103	0.418	0.101	0.424	0.0982	0.500	0.118
	(0.389)	(0.101)	(0.390)	(0.100)	(0.392)	(0.0967)	(0.421)	(0.106)
Age30DecilePI	−1.089*	−0.195***	−1.123**	−0.198***	−1.227**	−0.207***	−1.273**	−0.213***
	(0.426)	(0.0590)	(0.428)	(0.0579)	(0.442)	(0.0555)	(0.414)	(0.0521)
Age40DecilePI	−0.114	−0.0250	−0.145	−0.0314	−0.207	−0.0431	−0.208	−0.0437
	(0.388)	(0.0842)	(0.390)	(0.0833)	(0.396)	(0.0807)	(0.370)	(0.0759)
Age50DecilePI	−0.962*	−0.192**	−1.002*	−0.197**	−1.099**	−0.203**	−1.178**	−0.221***
	(0.395)	(0.0690)	(0.399)	(0.0682)	(0.414)	(0.0639)	(0.379)	(0.0619)
AmerPI	−0.390	−0.0897	−0.392	−0.0895	−0.362	−0.0795	−0.349	−0.0774

(continued)

Table 8.13 (continued)

	(1)		(2)		(3)		(4)	
	Failure	Marginal effects	*Failure*	Marginal effects	*Failure*	Marginal effects	*Failure*	Marginal effects
	(0.264)	(0.0610)	(0.265)	(0.0608)	(0.275)	(0.0606)	(0.264)	(0.0589)
Employees	0.00280	0.000624	0.00290	0.000640	0.00304	0.000647	0.00341	0.000732
	(0.00243)	(0.000522)	(0.00239)	(0.000509)	(0.00239)	(0.000491)	(0.00263)	(0.000540)
SimAwardsDummy	−1.052***	−0.252***	−1.069***	−0.255***	−1.052***	−0.240***	−1.063***	−0.245***
	(0.254)	(0.0574)	(0.258)	(0.0576)	(0.261)	(0.0559)	(0.261)	(0.0566)
ProfInvolved	−0.786*	−0.175*	−0.779*	−0.172*	−0.778*	−0.166*	−0.791*	−0.170*
	(0.352)	(0.0730)	(0.348)	(0.0718)	(0.359)	(0.0711)	(0.352)	(0.0699)
DMT	0.146	0.0324	0.181	0.0397	0.189	0.0400	0.232	0.0492
	(0.258)	(0.0570)	(0.265)	(0.0579)	(0.281)	(0.0591)	(0.265)	(0.0560)
Intercept	0.332		0.333		0.389		0.493	
	(0.417)		(0.413)		(0.440)		(0.407)	
Wald χ^2	30.44**		30.51**		33.64**		40.20***	
Likelihood ratio	41.1***		41.96***		46.36***		45.59***	

Robust standard errors in parentheses
*$p < 0.05$ ** $p < 0.01$ *** $p < 0.001$

Table 8.14 PI as sole founder and CEO *Failure* probit

	(1)		(2)	
	Failure	Marginal effects	*Failure*	Marginal effects
PIOneFounderCEO	1.018*	0.261*	0.863	0.276
	(0.397)	(0.109)	(0.465)	(0.183)
AcademicBackground			−0.173	−0.0290
			(0.140)	(0.0356)
PIOneFounderCEO × *AcademicBackground*			0.220	
			(0.735)	
FemalePI	−0.657	−0.116	−0.761	−0.129
	(0.649)	(0.0900)	(0.701)	(0.0883)
MinorityPI	0.424	0.0990	0.564	0.132
	(0.389)	(0.0974)	(0.401)	(0.101)
Age30DecilePI	−1.171**	−0.200***	−1.350**	−0.219***
	(0.431)	(0.0562)	(0.413)	(0.0501)
Age40DecilePI	−0.130	−0.0275	−0.227	−0.0469
	(0.388)	(0.0806)	(0.369)	(0.0750)
Age50DecilePI	−1.072**	−0.200**	−1.215**	−0.219***
	(0.401)	(0.0644)	(0.375)	(0.0590)
AmerPI	−0.360	−0.0796	−0.321	−0.0697
	(0.271)	(0.0604)	(0.269)	(0.0590)
Employees	0.00301	0.000643	0.00362	0.000766
	(0.00231)	(0.000477)	(0.00256)	(0.000517)
SimAwardsDummy	−1.050***	−0.241***	−1.055***	−0.238***
	(0.254)	(0.0557)	(0.258)	(0.0544)
ProfInvolved	−0.771*	−0.165*	−0.765*	−0.162*
	(0.352)	(0.0699)	(0.367)	(0.0717)
DMT	0.170	0.0360	0.183	0.0383
	(0.268)	(0.0567)	(0.262)	(0.0548)
Intercept	0.307		0.526	
	(0.412)		(0.405)	
Wald χ^2	31.60***		33.92**	
Likelihood ratio	45.78***		47.94***	

Robust standard errors in parentheses
$^*p < 0.05$ $^{**}p < 0.01$ $^{***}p < 0.001$

PIOneFounderCEO have a 26.1 percentage point increase in the probability of *Failure* over firms that do not. This result is consistent with the hypothesis that the more concentrated the experience of the firm leadership, i.e., the more roles a single person plays, the more likely the firm will experience *Failure.* The second estimation results shown in Table 8.14 augment the first with the inclusion of *AcademicBackground* as well as its interaction with *PIOneFounderCEO*. After controlling for *AcademicBackground,* the parameter estimate for *PIOneFounderCEO* is no longer significant at the 0.05 level. However, with a p value of 0.063, the estimate is suggestive of a significant relationship. The marginal effect of *PIOneFounderCEO* is also no longer significant at the 0.05 level. The parameter estimates for *AcademicBackground* and the interaction term are not statistically significant at a reasonable confidence level.

The next sets of model results use *TechFailure* as the dependent variable. Table 8.15 presents the estimation results of two models: the first includes *PIFounder* and the second *PIOneFounder*. As shown in the first specification results presented in Table 8.15, neither the parameter estimate nor the marginal effect for *PIFounder* are statistically significant. However, after controlling for this type of leadership experience, the coefficient for *Employees* is negative and statistically significant at the 0.05 level, whereas in the initial estimates, it was not significant at the reported alpha levels (see Table 7.7). The marginal effect is not significant at the 0.05 level, however, has a p value of 0.065, which may suggest a significant relationship. The

Table 8.15 PI as founder and as sole founder *TechFailure* probit

	(1)		(2)	
	TechFailure	Marginal effects	*TechFailure*	Marginal effects
PIFounder	−0.338	−0.0353		
	(0.343)	(0.0323)		
PIOneFounder			0.327	0.0422
			(0.391)	(0.0587)
MinorityPI	0.528	0.0755	0.566	0.0817
	(0.435)	(0.0742)	(0.442)	(0.0771)
Age30DecilePI	−0.286	−0.0299	−0.307	−0.0319
	(0.477)	(0.0459)	(0.482)	(0.0461)
Age40DecilePI	0.0321	0.00365	−0.00563	−0.000636
	(0.459)	(0.0524)	(0.457)	(0.0516)
Age50DecilePI	−0.740	−0.0687	−0.901	−0.0801*
	(0.568)	(0.0415)	(0.562)	(0.0406)
AmerPI	−0.302	−0.0354	−0.304	−0.0354
	(0.303)	(0.0353)	(0.307)	(0.0357)
Employees	−0.00751*	−0.000850	−0.00491	−0.000555
	(0.00329)	(0.000461)	(0.00259)	(0.000342)
SimAwardsDummy	−0.689*	−0.0798*	−0.781*	−0.0904**
	(0.309)	(0.0338)	(0.317)	(0.0348)
ProfInvolved	−0.539	−0.0610	−0.491	−0.0555
	(0.453)	(0.0514)	(0.442)	(0.0488)
DMT	−0.0119	−0.00135	0.0795	0.00904
	(0.277)	(0.0314)	(0.294)	(0.0337)
Intercept	−0.429		−0.609	
	(0.514)		(0.516)	
Wald χ^2	17.02		17.82	
Likelihood ratio	16.91		16.75	

Robust standard errors in parentheses
*$p < 0.05$ ** $p < 0.01$ *** $p < 0.001$

marginal effect of *Employees* on *TechFailure* is relatively small at −0.000850. This indicates that for each additional employee, the probability of *TechFailure* is reduced by 0.0850 percentage points. While the marginal effect of *Employees* on *TechFailure* is relatively small after controlling for *PIFounder,* it is directionally intuitive. In the case when firm leadership experience is less varied, the experience gained from adding additional human capital to the firm offsets the lack of variation in the experience of the firm leaders.

The second estimation results shown in Table 8.15 include *PIOneFounder* as the measure of firm leadership experience. In this specification, neither the parameter estimate nor the marginal effect of *PIOneFounder* are statistically significant. The parameter estimate for *Employees* is negative and has a *p* value of 0.058, which may suggest a significant relationship with *TechFailure*; however, the marginal effect of *Employees* on *TechFailure* in this model is not statistically significant.

The final set of estimation results is presented in Table 8.16 and augments those presented in Table 8.15 with further conditioning on PIs who were also CEOs. As shown in the first set of results in Table 8.16, the parameter estimate for *PICEO* and its marginal effect are not statistically significant. However, similar to the results in Table 8.15, the parameter estimate for *Employees* is negative and significant at the 0.05 level. The marginal effect is −0.000714 and has a *p* value of 0.079 suggesting a small effect on *TechFailure*.

The second estimation results shown in Table 8.16 include an even more concentrated firm leadership structure with the use of *PIFounderCEO.* Although *PIFounderCEO* is not a statistically significant estimator of *TechFailure*, after controlling for this variable, the coefficient on *Employees* is significant at the 0.05 level. The *p* value of the marginal effect of *Employees* on *TechFailure* is 0.078. This marginal effect indicates that an additional employee reduces the probability of *TechFailure* by 0.0692 percentage points.

The last model estimation results presented in Table 8.16 include the most concentrated firm leadership experience variable, *PIOneFounderCEO.* Similar to the prior two specifications, the coefficient and marginal effect for *PIOneFounderCEO* are not statistically significant.

In summary, the findings of the analysis of firm leadership experience on project failure suggest that the more concentrated the human capital within the firm leadership, the more likely the project will fail. Further, the vocational background of the firm's leaders may have an impact on the probability of *Failure*. The human capital gained by firm leaders from having a business background or working in academia may translate to experience useful in reducing *Failure.* The concentration of firm leadership experience does not have a meaningful impact on failure for technical reasons. However, after considering the concentration of human capital of firm leaders, having more employees is a suggestive indicator for a reduction in the probability of *TechFailure.* By increasing the number of employees, the experience base of human capital is increased, which may help with preventing issues that lead to technical failures.

Finally, variance inflation factors (VIFs) for the group of firm and project characteristics that were used in the prior sets of estimations of Eq. (7.1) (Tables 8.5, 8.6,

Table 8.16 PI CEO and CEO founder *TechFailure* probit

	(1)		(2)		(2)	
	TechFailure	Marginal effects	*TechFailure*	Marginal effects	*TechFailure*	Marginal effects
PICEO	−0.0868	−0.00964				
	(0.357)	(0.0382)				
PIFounderCEO			−0.0478	−0.00537		
			(0.361)	(0.0397)		
PIOneFounderCEO					0.483	0.0662
					(0.407)	(0.0694)
MinorityPI	0.535	0.0772	0.536	0.0775	0.579	0.0832
	(0.434)	(0.0749)	(0.434)	(0.0751)	(0.449)	(0.0780)
Age30DecilePI	−0.291	−0.0306	−0.291	−0.0306	−0.349	−0.0356
	(0.474)	(0.0458)	(0.471)	(0.0455)	(0.478)	(0.0447)
Age40DecilePI	0.0139	0.00158	0.0107	0.00123	−0.0355	−0.00397
	(0.454)	(0.0520)	(0.455)	(0.0521)	(0.452)	(0.0503)
Age50DecilePI	−0.792	−0.0730	−0.802	−0.0738	−0.973	−0.0843*
	(0.559)	(0.0409)	(0.557)	(0.0408)	(0.562)	(0.0398)
AmerPI	−0.306	−0.0360	−0.305	−0.0359	−0.292	−0.0338
	(0.304)	(0.0357)	(0.304)	(0.0356)	(0.312)	(0.0362)
Employees	−0.00626*	−0.000714	−0.00607*	−0.000692	−0.00454	−0.000510
	(0.00296)	(0.000406)	(0.00287)	(0.000393)	(0.00254)	(0.000329)
SimAwardsDummy	−0.725*	−0.0847*	−0.732*	−0.0856*	−0.801*	−0.0918**
	(0.314)	(0.0347)	(0.314)	(0.0348)	(0.324)	(0.0351)
ProfInvolved	−0.546	−0.0623	−0.543	−0.0619	−0.486	−0.0546
	(0.442)	(0.0499)	(0.441)	(0.0497)	(0.444)	(0.0487)
DMT	0.0466	0.00534	0.0515	0.00590	0.102	0.0116
	(0.280)	(0.0322)	(0.279)	(0.0321)	(0.289)	(0.0332)
Intercept	−0.523		−0.534		−0.615	
	(0.507)		(0.504)		(0.506)	
Wald χ^2	16.73		16.61		18.15	
Likelihood ratio	16.3		16.28		17.22	

Robust standard errors in parentheses
$^{*}p < 0.05$ $^{**}p < 0.01$ $^{***}p < 0.001$

8.7, 8.11, 8.12, 8.13, 8.14, 8.15, and 8.16) are shown in Table 8.17. As shown in Table 8.18, the VIF on each variable is less than 2 except for the age decile indicators, which are less than 3; this suggests that there is little concern for multicollinearity between these variables. Since these sets of variables encompass all the variables used in the models described in this chapter, multicollinearity is not a concern for any of the prior estimations of Eq. (7.1).

Table 8.17 Variance inflation factors

	Table.Model_Number						
	8.11.2	8.11.3	8.12.2	8.13.1	8.13.3	8.13.4	8.14.2
PIFounder	1.7982	1.8035					
PIOneFounder			1.7744				
PICEO				1.1734			
PIFounderCEO					1.7245	1.6681	
PIOneFounderCEO							1.7222
Bizbackground	1.5276				1.4272		
AcademicBackground		1.4098	1.1312			1.2963	1.1187
FemalePI	1.0584	1.0610	1.0756	1.0473	1.0539	1.0611	1.0737
MinorityPI	1.3229	1.3297	1.3233	1.3177	1.3216	1.3407	1.3230
Age30DecilePI	2.3070	2.2995	2.2382	2.2027	2.2669	2.2446	2.2368
Age40DecilePI	2.5942	2.4972	2.4717	2.4744	2.6151	2.5170	2.4838
Age50DecilePI	2.4676	2.4484	2.4651	2.4390	2.5152	2.4863	2.4914
AmerPI	1.3351	1.3212	1.3187	1.3205	1.3376	1.3208	1.3250
Employees	1.1635	1.1803	1.1233	1.1601	1.1588	1.1738	1.1269
SimAwardsDummy	1.1200	1.1206	1.1143	1.1111	1.1172	1.1235	1.1099
ProfInvolved	1.0889	1.0814	1.1084	1.0770	1.0855	1.0792	1.0997
DMT	1.1516	1.1319	1.1035	1.0992	1.1290	1.1156	1.1103

Table 8.18 Mean values by gender composition (standard deviations in parenthesis)

Variable	Project composition			
	Female PI/female owner	Female PI/male owner	Male PI/female owner	Male PI/male owner
Failure	0	0.17	0.14	0.22
TechFailure	0	0	0	0.08
SimAwards	0.67	0.50	0.86	1.22
	(0.58)	(0.55)	(0.90)	(1.82)
Employees	9.67	53.17	27.71	36.36
	(5.77)	(57.13)	(49.79)	(49.31)
n	3	6	7	153

8.4 Homophilic Gender Composition

The final PI and firm relationship considered here is the case when both the PI and firm owner are of the same gender. Bednar et al. (2019) found a statistically significant positive relationship between the probability of commercializing a technology and if the firm owner and PI were both females. In this light, the homophilic relationship between owner and PI is analyzed here using the data from the random sample of 169 Phase II–funded projects discussed previously. Due to the lack of variation in the data, i.e., no female owner/female PI firms experienced *Failure*, the reduced form model described in Chap. 7 is not able to be leveraged. However, what is presented are insights drawn from the data. Table 8.18 presents the summary statistics of the project team and project outcomes by gender composition of the project team.

As shown in the table, female-owned firms experienced *Failure* at a lower mean rate than their male counterparts. Firms with a female PI and a female owner ($n = 3$) did not have any *Failure*, while the firms with a female PI and a male owner ($n = 6$) had a mean *Failure* of 0.17. Using the count of similar awards for each firm, *SimAwards*, the female PI/female owner firms received slightly more *SimAwards*, at the mean, 0.67, than female PI/male owner firms, which received a mean of 0.5 *SimAwards*. Female PI/female owner firms were much smaller, at the mean, with close to ten employees, than female PI/male owner firms, which had the largest number of employees on average with a mean of about 53 *Employees*. Using *SimAwards* as a measure of experience and *Employees* as a measure of project complexity, the female PI/female owner firms were slightly more experienced than the female PI/male owner firms; however, the female PI/female owner firms are much smaller, which may indicate the projects pursued by the female PI/female owner firms were less complex than the female PI/male owner projects. From these results, it may be inferred that female PI/female owner firms are more likely to take on small, perhaps less risky projects than female PI/male owner firms. It is commonly accepted that females tend to be more risk-averse than their male counterparts, on average. Therefore, the homophilic female PI/female owner relationship experiences *Failure* at a much lower rate than the female PI/male owner firms, but the projects are also likely less risky.

Further, male PI/male owner firms ($n = 153$) experienced *Failure* at the highest rate (0.22) among the four gender combinations. Male PI/male owner firms had the largest mean *SimAwards* of 1.22, indicating that firms with this gender composition tend to have more experience than others, however, also tend to experience *Failure* more often. The mean number of *Employees* of the male PI/male owner firms is about 36, the second largest average firm size. Again, using firm size as a measure of project complexity, male PI/male owner firms typically engage in projects riskier than those of female owners with either male or female PI, but less risky than female PI/male owner firms. Thus, female PI/male owner firms may tend to undertake larger, riskier projects as measured by *Employees* and on average have less experience as measured by *SimAwards* than male PI/male owner firms; however, male PI/male owner firms still have a higher mean *Failure*.

Male PI/female owner firms ($n = 7$) experience *Failure* at a mean rate of 0.14, which is just 3 percentage points lower than firms with a female PI/male owner composition. Male PI/female owner firms received a mean of 0.86 *SimAwards* compared to a mean of 0.5 *SimAwards* received by female PI/male owner firms. Male PI/female owner firms had a mean firm size of almost 28 employees, which is most similar to male PI/male owner firms. Therefore, male PI/female owner firms may have a similar failure rate as female PI/male owner firms as these firms are typically smaller and slightly more experienced.

As shown, both homophilic compositions of firms hold the two boundaries of *Failure*. Female PI/female owner firms have the lowest mean rate of *Failure*, while male PI/male owner firms have the highest mean rate of *Failure*. Although the female homophilic firm is typically less experienced than their male counterparts, they tend to have smaller firms, indicating the lower *Failure* rate can at least be partially explained by risk aversion. Further, conditional on firms with female owners, firms with a female PI are less experienced, have smaller firm sizes, and have lower rates of *Failure* on average. Conditioning on firms with male owners, firms with a female PI are typically less experienced but have larger firm sizes, however, a lower mean rate of *Failure*, compared to firms with a male PI. All this together provides further evidence of a negative association between *FemalePI* and *Failure*, and evidence that the homophilic relationship of *FemalePI/Female Owner* firms tend to experience lower rates of *Failure* than any other gender combination of firm owner and PI.

8.5 Conclusion

This chapter presented analyses that augmented the firm and project characteristics discussed in Chap. 7 with information on PIs that have yet to be presented in the literature. Demographic information on PIs was added to the model specification that already included firm and project characteristics to test for statistically significant associations with SBIR project failure. From this analysis, it has been shown that there is a statistically significant association between the age of the PI and the likelihood of project failure. Specifically, younger cohorts of PIs were estimated to experience failure at lower rates relative to the oldest cohort of PIs. This result has two intuitive explanations. First, it is possible that on average, the projects that younger PIs work on are less risky and therefore less likely to fail. Second, the oldest cohort of PIs may be suffering from decreasing marginal human capital as the length of their remaining career diminishes.

Additionally, this analysis has shown that there is a statistically significant, positive relationship between the concentration of the firm leadership by an individual and SBIR project failure. The most concentrated measure of leadership in this analysis is when a PI is also the CEO and the sole founder of the firm. When an individual holds these three roles, the experience and human capital of the firm's leadership are concentrated on that single individual. This concentration results in the largest likelihood of project failure relative to the other factors considered. When an individual has too much responsibility or too many duties to perform, especially duties critical to the firm or project, then the success of that project and/or firm will suffer. Additionally, as shown previously, there is a negative relationship between R&D project failure and having a university professor involved in the R&D project likely due to the additional human capital a professor adds. Similarly, firms benefit from the additional human capital when the leadership of the firm is not concentrated on one individual.

In the next and final chapter, a concluding discussion is provided along with a description of the limitations associated with the analysis performed. Thoughts on policy implications as a result of the findings discussed previously are detailed, and ideas for future research are given.

Part II
Reflections and Future Perspectives

Chapter 9
PIs and R&D Project Failure: Reflections and Future Research Avenues

Abstract The concluding chapter of this monograph begins with a discussion that summarizes how the literature review on principal investigators (PIs) and research and development (R&D) project failure, the theoretical model of project failure, and the empirical analysis introducing PIs as a covariate with R&D project failure contributes to the existing literature. A recap of the estimation results of Small Business Innovation Research (SBIR) project failure is provided that gives a summary of the empirical findings and an interpretation of their relevance. Following the discussion of the empirical results, the known limitations of the study are outlined, and a discussion of the severity of the limitations is provided. As a mitigating factor to some of the limitations, future research ideas are shared. Additionally, future ways to deepen this line of research are also provided beyond those inspired by the known limitations. For example, the theoretical model could be further developed by introducing a term that reflects the marginal costs of experience. Two measures of failure are considered in this analysis; however, other measures of failure could be useful to explore. Finally, in the closing of this chapter, a brief final summary of the work provides a high-level recap as to what was discussed and the overall relevance of the monograph.

9.1 Introduction

This chapter serves as a summary of the most relevant information provided in the previous chapters. First, a discussion is provided that highlights the Department of Energy (DOE) and its involvement in the Small Business Innovation Research (SBIR) program. A brief review of the role principal investigators (PIs) play in research and development (R&D) project failure and results of the empirical analysis are provided.

Following the summary, a perspective of the potential implications to policy and practice, given the results of this analysis, is offered. Any study, at least in the field of economics, is a model of a broader domain and, therefore, has limitations. Known

M. Boyce, *Principal Investigators and R&D Failure*, Academic Entrepreneurship, Scientists, and Scientific Careers, https://doi.org/10.1007/978-3-031-43608-6_9

limitations of this study are provided followed by future research avenues of which some could compensate for the limitations discussed. To conclude this chapter, a brief synopsis of the relevance of this monograph to the literature and a final reflection of empirical findings are provided.

9.2 Discussion

9.2.1 *DOE's SBIR Program*

The DOE has a long-standing history of using technological innovation to achieve its goals. As discussed previously, the DOE descended from one of the best-known projects that involved public–private partnerships resulting in a new technology coming to fruition. Given the current size, and therefore budget of the DOE, the agency is one of the top contributors to the SBIR program. The SBIR program is the main source of public funding in the United States for small firms conducting innovative research with the goal of commercializing new technologies. Each firm that is funded through the SBIR program should have a principal investigator acting as the lead scientist for the project. As discussed previously, the role of the PI is critical in small innovative firms. Although a PI may be the lead scientist for an R&D project, a PI's responsibilities can span beyond the typical day-to-day responsibilities of a research scientist and encompass duties akin to those of a manager or leader of the firm. Given the relative importance of PIs in the innovation process, it is important to understand what characteristics of PIs, if any, are associated with a greater likelihood of project failure.

9.2.2 *R&D Project Failure and the Role of PIs*

The literature examining the determinants of SBIR project failure is limited, and there are no studies that examine the breadth of characteristics of PIs and their association with failure as is done in this book. As such, this book provides several important contributions to the literature.

A complete literature review on R&D project failure that spans both the economics and management disciplines literature was conducted to understand past research that could be used as a stepping stone for the analysis in this monograph. There is a notable lack of research that considers PIs or more generally an R&D project leader as a source of project failure. Further, there is no theoretical model for research project failure in the existing literature. To fill this void, this monograph provides a new-to-the-literature theory that uses a structural form model to explain how firm and project characteristics, such as PI's experience, may impact the likelihood of R&D project failure. The theory is struck in terms of entrepreneurial firms conducting R&D with the intent of earning a profit; however, it is generalizable to other firm types.

9.2.3 Review of Empirical Results

Using the DOE SBIR data from the National Research Council (NRC) second round survey, which has not been described in the literature outside of this book, a set of initial models were estimated using a reduced form specification of the structural model. The results of these estimations provide support for two of the most relevant previous studies' findings. Given the results from Andersen et al. (2017) and Link and Wright (2015), an a priori expectation that having received a previous award would be negatively associated with project failure was anticipated. Having received previous awards is a measure of experience, and that experience is shown to be beneficial for reducing the probability of project failure. This result is intuitive because whether firms or PIs who received a previous award experienced failure or not, the experience gained from going through the R&D process, even partially, is a knowledge stepping stone for overcoming the pitfalls that may lead to project failure in future projects.

From the initial set of models, it was found that having university faculty involved in the R&D project translated to a relatively large reduction in the probability a project failed. This result is in line with the original expectation provided, which was motivated by the findings in Gicheva and Link (2016). Having university faculty involved in the project is a measure of human capital, and given the nature of their work, university faculty provide research expertise that tends to reduce the probability of project failure. Knowing the effect on the likelihood of project failure when university faculty are involved in a Phase II–funded SBIR project is relevant information to program applicants. In terms of policy, SBIR program administrators could use this information to educate applicants and encourage or potentially incentivize applicants to involve university faculty in their project.

After confirming results found in previous studies (Link & Wright, 2015; Gicheva & Link, 2016; Andersen et al., 2017), additional PI experience variables were added to the initial model's specification. Three general measures to describe PIs' experience were analyzed: demographic characteristics, PIs as firm leaders, and homophilic relationship between firm owners and PIs. From the demographic characteristics of PIs, using age decile bins, the estimations showed that younger PIs, those younger than the oldest cohort (PIs at least 60 years old), tend to have a lower probability of failure than the oldest PIs in the random sample. This result may be because younger PIs take on less complex projects that naturally are less likely to fail and because of higher levels of motivation driven by higher rates of marginal human capital accumulation. On the other hand, the oldest cohort of PIs may be less risk-averse than their younger counterparts but also experience depreciating rates of human capital. The oldest PIs may have enjoyed past success and are willing to take on riskier projects because their opportunity cost of failing is lower than their younger counterparts.

Cunningham et al. (2016a) found that PIs take on multiple roles such as project manager, administrator, science broker, and boundary spanner (i.e., ability to bridge different areas and domains such as the academic sector and the private sector). Through the empirical work discussed previously, it was found that in addition to these roles, PIs may also be a founder of the firm and/or chief executive officer (CEO). When PIs are founders and/or CEOs of the firm, the leadership structure is more concentrated, and the responsibilities of the PI are greater. In general, the more concentrated the leadership, the greater the probability a project will fail. The case when the PI is the sole founder and CEO has the largest marginal effect on the probability a firm would fail. This case is the most concentrated firm leadership structure in the random sample and suggests that although PIs may have the ability to take on many responsibilities, there is a limit to the number of roles a PI can assume and still have a project that does not fail. These results provide useful information that could be used to inform policy of the SBIR program. SBIR program administrators could inform award recipients, especially those who have a concentrated leadership structure, to consider creating a greater span of control by employing a CEO other than the PI. Providing firms with information that may reduce the aggregate likelihood of failure across the SBIR's award recipients will increase the return on public investment in the SBIR program.

Additionally, it was found that the vocational background of PIs who were a founder or both a founder and the CEO has an impact on the probability of project failure. For the firms where the PI was a founder and not the only founder, having founders with either a business background or academic background resulted in having a lower probability of project failure. Founders having a business or academic background also had a negative effect on project failure for firms that had a PI who was both a founder and the CEO. These results imply that the experience gained or skills obtained through specific vocational experience provide a benefit to firms in reducing the probability of failure when the firm's leadership structure is somewhat concentrated. However, for the firms with the most concentrated leadership structure, the case when the PI was the sole founder and the CEO of the firm, the vocational background of the PI did not have a significant impact on failure. In this case, the experience gained from being in business or academia does not offset the positive impact on failure resulting from the workload of being the sole founder, CEO, and PI.

9.3 Limitations

As with all studies, the analysis performed in this book has limitations. While the data used in this book are a random sample of Phase II–funded SBIR projects from the DOE's SBIR program, it would be useful to have a larger sample size. There is rarely, if ever, a case in an economic study where more data would not be useful.

Additionally, the data are from a single department's SBIR program, so the results may not be generalizable across all SBIR participating agencies. Having similar data from other SBIR participating agencies would increase the sample size and provide insight into the generality of the results discussed previously. There is a notable lack of information on the Phase II award amount each firm received, which has been shown to be a highly significant covariate with project failure (Link & Wright, 2015). Another measure that would be useful but not available is a variable that represents the cost of a PI's experience. It is straightforward to hypothesize that a better experience would come with a higher cost and a lower probability of failure, on average. Further, R&D projects have a time component; it takes time to move a project through different phases. However, the data used in this research do not provide information about the timing of the failure of the projects in the random sample. This information would be useful to further understand and define additional measures of failure and perhaps pinpoint specific phases in the R&D process that lead to the demise of a project. Finally, it was not possible to provide an estimate of the probability of failure with the female homophilic relationship between owner and PI as a covariate due to the lack of representation of this type of relationship in the random sample.

9.4 Implications for Policy and Practice

Given the results of the empirical analysis, recommendations for policy and/or practice follow. The National Academies program evaluations of the SBIR program at the DOE (Academies, 2016 and Academies, 2020) are two of the most comprehensive evaluations of the DOE's SBIR program. Both reviews offer recommendations of how the DOE can make improvements to its program to further align program outcomes to the stated goals of the SBIR program. We would like to reiterate some of those recommendations, however, with a more specific focus on PIs. Perhaps the most general recommendation comes from Academies (2016), which recommended the DOE improve its data collection and tracking mechanisms. Additional data such as a percent complete or percent success rate of an R&D project would be useful to understand nonbinary success/failure as well as a measure of time spent through key stages of the project. In a similar vein, Academies (2020, p. 9) recommends that "The DOE SBIR/STTR program office should develop better metrics of potential commercialization by applicants and commercialization outcomes by awardees." We recommend DOE-SBIR programs administrators adopt a framework motivated by the theoretical model and the empirical analysis provided in this monograph to aid in evaluating the likelihood of success of DOE-SBIR applicants projects. Additionally, it would be useful to implement a risk ranking system for SBIR projects as an additional measure to consider when evaluating project outcomes.

Regarding human capital at the firm, we recommend the DOE SBIR program office help prospective applicants identify university professors who have experience in the R&D project field to enlist as project members or advisors. This echoes recommendation 4.1 in Academies (2020, p. 99) which recommended the DOE SBIR/STTR (Small Business Technology Transfer) program office inform applicants of perspective R&D partners. Additionally, we recommend the DOE SBIR program office inform prospective applicants, when applicable, of the potential benefit of having different people in key leadership roles. The program office should also inform prospective applicants of other firms or professionals who may be fitting partners for reducing the risk of failure given a concentration of firm leaders.

Finally, in Academies (2016, p. 8), there is a recommendation to monitor the composite age of firm applicants to understand if the age of the firm effects SBIR outcomes. We suggest extending this monitoring to incorporate the age of the PI given the association between the age of the PI and the likelihood of SBIR project failure shown in Chap. 8.

9.5 Future Research

The analysis performed in this book as well as the limitations associated with it gives rise to several avenues for future research. First, to test the robustness of the results presented earlier, future research may consider using additional data from other SBIR participating departments to conduct similar analyses to those performed here. Conditional on the variables available, leveraging the data used by Link and Wright (2015), Gicheva and Link (2016), Bray and Link (2017) would be an immediate consideration for additional data sources. It would also be useful to conduct a systematic literature review on determinants of R&D project failure. The systematic review may provide insight into potential data sources to further test the robustness of the results found in this monograph.

The empirical work in this piece considered two measures of failure. However, framing failure across different measures would be interesting. For example, failure could be defined as not receiving a patent, not generating spinoffs, or not increasing employment over the duration of the project. Examining additional specific types of failure beyond technical failure such as market failure or operational failure may also provide interesting insights. Some of these measures could be formulated using the data analyzed in this book. Nonetheless, studying different objectives of the firm may help uncover useful information or covariates that are beneficial for reducing failure across several objectives or perhaps covariates unique to a specific measure of failure.

Future research may look at the specific time or phase within the R&D project life cycle that the project failed, and this concept could potentially be incorporated to advance the theoretical model. With this information, policymakers could inform program participants of the pitfalls associated with various phases of an R&D project and perhaps channel resources to help generate an overall, SBIR-wide, lower probability of failure. Further, another expansion to the theoretical model could be to incorporate an additional measure for the cost of the experience of the principal investigator. On average, the better the experience a principal investigator has accumulated, the more that principal investigator would cost since better experience should, at least partially, reduce the risk of failure. It would also be interesting to enhance the model to consider the expected outcome with and without public funding. In this case, the model could be reformulated to consider expected return on investment as a determinant of project success/failure.

Additionally, understanding or formulating a measure of risk associated with SBIR projects would provide another useful dimension to understand why projects fail or even a method for categorizing SBIR projects. Knowing the variation in risk between projects may also provide further insights into characteristics of principal investigators who are associated with projects that bear certain risk profiles. One measure that could help in defining the risk of a project is the SBIR award amount. Since awards are given to help further the innovation, that is to move toward commercialization, it is not known whether the project will succeed or fail when the award is given. Therefore, the amount of the award could be considered as a measure of risk, whereas the greater monetary value of the award translates to greater potential losses, hence greater risk.

9.6 Conclusion

In conclusion, this monograph provides a complete literature review on R&D project failure that spans both the economics and management disciplines literature. This review uncovered a notable lack of literature related to PIs and their association with project failure as well as the absence of a theoretical model to explain, mathematically, research failure. This book provides a theoretical model of research project failure to help fill the void, which is used in its structural form, to guide the empirical analysis conducted. Further, an analysis of DOE data using NRC second-round survey data was provided, and these data were used to estimate an empirical probability of failure, which resulted in findings that provide support for previous studies' findings. This study presents PIs as a new-to-the-literature covariate with R&D project failure and found that certain characteristics of PIs lead to a lower probability of failure. Specifically, younger PIs have a lower probability of failure than the oldest cohort, and PIs who are the sole founders of the firm and are also the CEO have a significantly higher probability of failure compared to PIs who are strictly PIs. As such, this book contributes to a small but growing body of literature that considers the role PIs have in the failure of R&D projects.

References

Acs, Z. J., & Audretsch, D. B. (1988). Innovation in large and small firms: An empirical analysis. *The American Economic Review, 78*(4), 678–690.

Allen, S. D., Layson, S. K., & Link, A. N. (2012). Public gains from entrepreneurial research: Inferences about the economic value of public support of the small business innovation research program. *Research Evaluation, 21*, 105–112.

Andersen, M. S., Bray, J. W., & Link, A. N. (2017). On the failure of scientific research: An analysis of SBIR projects funded by the U.S. National Institutes of Health. *Scientomerics, 112*(1), 431–442.

Audretsch, D. B., Link, A. N., & Scott, J. T. (2002). Public/private technology partnerships: Evaluating SBIR-supported research. *Research Policy, 31*(1), 145–158.

Audretsch, D. B., Leyden, D. P., & Link, A. N. (2012). Universities as research partners in publicly supported entrepreneurial firms. *Economics of Innovation and New Technology, 21*(5–6), 529–545.

Baglieri, D., & Lorenzoni, G. (2014). Closing the distance between academia and market: Experimentation and user entrepreneurial processes. *The Journal of Technology Transfer, 39*(1), 52–74.

Bednar, S., Gicheva, D., & Link, A. N. (2019). Innovative activity and gender dynamics. *Small Business Economics*. https://doi.org/10.1007/s11187-019-00282-2

Bhattacharya, V. (2018). An empirical model of R&D procurement contests: An analysis of the DOD SBIR program. (Online Appendix). Revision requested, *Econometrica*.

Birch, D. G. W. (1979). The job generation process: MIT program on neighborhood and regional. *Change, 302*.

Boehm, D. N., & Hogan, T. (2014). 'A jack of all trades': The role of PIs in the establishment and management of collaborative networks in scientific knowledge commercialisation. *Journal of Technology Transfer, 39*(1), 134–149.

Bray, J. W., & Link, A. N. (2017). Dynamic entrepreneurship: On the performance of U.S. research joint ventures. *Small Business Economics, 49*(4), 785–797. https://doi.org/10.1007/s11187-017-9862-x

Bryan, K., & Williams, H. (2021). Innovation: Market failures and public policies. NBER Working Paper No. 29173.

Carter, J. (1977). *An address to the nation on energy. Online by Gerhard Peters and John T.* Woolley.

Cullison, W. E. (1989). The U.S. productivity slowdown: What the experts say. *Federal Reserve Bank of Richmond: Economic Review, 75*, 10–21.

M. Boyce, *Principal Investigators and R&D Failure*, Academic Entrepreneurship, Scientists, and Scientific Careers, https://doi.org/10.1007/978-3-031-43608-6

Cunningham, J. A., O'Reilly, P., O'Kane, C., & Mangematin, V. (2014). The inhibiting factors that principal investigators experience in leading publicly funded research. *The Journal of Technology Transfer, 39*(1), 93–110.

Cunningham, J. A., O'Reilly, P., O'Kane, C., & Mangematin, V. (2016a). Publicly funded principal investigators as transformative agents of public sector entrepreneurship. In D. Audretsch & A. N. Link (Eds.), *Essays in public sector entrepreneurship* (pp. 67–94). Springer.

Cunningham, J. A., Mangematin, V., O'Kane, C., & O'Reilly, P. (2016b). At the frontiers of scientific advancement: The factors that influence scientists to become or choose to become publicly funded principal investigators. *The Journal of Technology Transfer, 41*(4), 778–797.

Cunningham, J. A., O'Reilly, P., Dolan, B., O'Kane, C., & Mangematin, V. (2016c). Publicly funded principal investigators allocation of time for public sector entrepreneurship activities. Economia E Politica Industriale. *Journal of Industrial and Business Economics, 43*(4), 383–408.

Cunningham, J. A., Menter, M., & Wirsching, K. (2017). Entrepreneurial ecosystem governance: A principal investigator-centered governance framework. *Small Business Economics.* https://doi.org/10.1007/s11187-017-9959-2

Cunningham, J. A., O'Reilly, P., Hooper, D., Nepelski, D., & Roy, V. V. (2020). *The role of project coordinators in European Commission framework programme projects.* Joint Research Centre.

Del, G. M., Nicotra, M., Romano, M., & Schillaci, C. E. (2017). Entrepreneurial performance of principal investigators and country culture: Relations and influences. *The Journal of Technology Transfer, 42*(2), 320–337.

Department of Energy. (2014). *Strategic plan: 2014–2018.* Retrieved from https://www.energy.gov/sites/prod/files/2014/04/f14/2014_dept_energy_strategic_plan.pdf

Department of Energy. (2015). *Office of science user facilities.* Retrieved from https://science.energy.gov/~/media/_/pdf/user-facilities/Reports/DOE-SC-User-Facilities-FY2015-report.pdf

Department of Energy. (n.d.). *About us.* Retrieved from https://www.energy.gov/about-us

Department of Energy—Office of History and Heritage. (n.d.). *Manhattan project: A tentative decision to build the Bomb.* Retrieved from https://www.osti.gov/opennet/manhattan-project-history/Events/1939-1942/tentative_decision_build.htm

Department of Energy—Office of Science. (n.d.-a). *About.* Retrieved from https://science.energy.gov/about/

Department of Energy—Office of Science. (n.d.-b). *Small Business Innovation Research (SBIR) and Small Business Technology Transfer (STTR).* Retrieved from https://science.osti.gov/sbir

Department of Energy—Office of SBIR/ STTR Programs. (2022). *Small Business Innovation Research (SBIR) and Small Business Technology Transfer (STTR) Programs.* Retrieved from https://science.osti.gov/-/media/sbir/powerpoint/2023/FY23-PIR1-FOA-Webinar_ZA_EEC.pptx

Department of Energy—Office of SBIR/ STTR Programs. (n.d.). *Program areas.* Retrieved from https://www.energy.gov/science/sbir/small-business-innovation-research-and-small-business-technology-transfer

Environmental Protection Agency. (2018, July 9). *Summary of the atomic energy act.* Retrieved from https://www.epa.gov/laws-regulations/summary-atomic-energy-act.

Fehner, T. R., & Holl, J. M. (1994, November). *Department of energy 1977–1994: A summary history. DOE/HR-0098.* Department of Energy, 1994. 142pp. History Office publication.

Feldman, M., Hadjimichael, T., Lanahan, L., & Kemeny, T. (2016). The logic of economic development: A definition and model for investment. *Environment and Planning. C, Government & Policy, 34*(1), 5–21. https://doi.org/10.1177/0263774X15614653

Fichtner, J. J., & Michel, A. N. (2015). *Can a research and development tax credit be properly designed for economic efficiency?* Mercatus Research, Mercatus Center at George Mason University.

Geunaa, A., & Nesta, L. J. J. (2006). University patenting and its effects on academic research: The emerging European evidence. *Research Policy, 35*, 790–807.

Gicheva, D., & Link, A. N. (2013). Leveraging entrepreneurship through private investments: Does gender matter? *Small Business Economics, 40*(2), 199–210.

Gicheva, D., & Link, A. N. (2016). On the economic performance of nascent entrepreneurs. *European Economic Review, 86*, 109–117.

Gordon, I. R., & P. McCann. (2005). Innovation, agglomeration, and regional development. *Journal of Economic Geography, 5*(5), 523–543.

Hall, B., & Reenen, J. V. (2000). How effective are fiscal incentives for R&D? A review of the evidence. *Research Policy, 29*, 449–469.

Hayter, C. S., Link, A. N., & Scott, J. T. (2018). Public-sector entrepreneurship. *Oxford Review of Economic Policy, 34*(4), 676–694. https://doi.org/10.1093/oxrep/gry014

Kidwell, D. K. (2013). Principal investigators as knowledge brokers: A multiple case study of the creative actions of PIs in entrepreneurial science. *Technological Forecasting and Social Change, 80*(2013), 212–220.

Kobe, K., & Schwinn, R. (2018). *Small Business GDP 1998–2014*. Retrieved from https://cdn.advocacy.sba.gov/wp-content/uploads/2018/12/21060437/Small-Business-GDP-1998-2014.pdf

Leyden, D. P., & Link, A. N. (2015). *Public sector entrepreneurship: US technology and innovation policy*. Oxford University Press.

Leyden, D. P., Link, A. N., & Siegel, D. S. (2014). A theoretical analysis of the role of social networks in entrepreneurship. *Research Policy, 43*(7), 1157–1163. https://doi.org/10.1016/j.respol.2014.04.010

Link, A. N., & Ruhm, C. J. (2009a). *Public knowledge, private knowledge: The intellectual capital of entrepreneurs*. NBER Working Papers 14797, National Bureau of Economic Research, Inc.

Link, A. N., & Scott, J. T. (2000). Estimates of the social returns to small business innovation research projects. In: C. W. Wessner (Ed.), *The small business innovation research program* (pp. 275–290). National Academy Press.

Link, A. N., & Ruhm, C. J. (2009b). Bringing science to market: Commercializing from NIH SBIR awards. *Economics of Innovation and New Technology, 18*, 381–402.

Link, A. N., & Scott, J. T. (2009). Private investor participation and commercialization rates for government-sponsored research and development: Would a prediction market improve the performance of the SBIR Programme? *Economica, 76*(302), 264–281.

Link, A. N., & Scott, J. T. (2010). Government as entrepreneur: Evaluating the commercialization success of SBIR projects. *Research Policy, 39*(5), 589–601.

Link, A. N., & Scott, J. T. (2012a). Employment growth from public support of innovation in small firms. *Economics of Innovation and New Technology, 21*(7), 655–678.

Link, A. N., & Scott, J. T. (2012b). Employment growth from the small business innovation research program. *Small Business Economics, 39*(2), 265–287.

Link, A. N., & Scott, J. T. (2013). *Bending the arc of innovation: Public support of R&D in small, entrepreneurial firms*. Palgrave Macmillan.

Link, A. N., & Wright, M. (2015). On the failure of R&D projects. *IEEE Transactions on Engineering Management, 62*(4), 442–448.

Link, A. N., Siegel, D. S., & Van, F. D. D. (2011). Public science and public innovation: Assessing the relationship between patenting at U.S. National laboratories and the Bayh-dole act. *Research Policy, 40*(8), 1094–1099.

Mangematin, V., O'Reilly, P., & Cunningham, J. (2014). PIs as boundary spanners, science and market shapers. *The Journal of Technology Transfer, 39*(1), 1–10.

Mansfield, E., & Wagner, S. (1975). Organizational and strategic factors associated with probabilities of success in industrial R&D. *The Journal of Business, 48*(2), 179–198.

Menter, M. (2016). Principal investigators and the commercialization of knowledge. In D. B. Audretsch, E. E. Lehmann, S. Vismara, & M. Meoli (Eds.), *University evolution, entrepreneurial activity and regional competitiveness* (pp. 193–203). Springer.

Mitchell III, W., & Turner, S. E. (1971). Breeder Reactors. Library of Congress Catalog Card Number: 76–611325 1971.

Mowery, D. C., & Sampat, B. N. (2004). The Bayh-dole act of 1980 and university–industry technology transfer: A model for other OECD governments? *The Journal of Technology Transfer, 30*(1/2), 115–127. https://doi.org/10.1007/s10961-004-4361-z

Murphy, L. M., Jerde, P., Rutherford, L., & Barone, R. (2007). *Enhancing commercial outcomes from R & D: A framework for a public-private partnership to increase the yield of federally funded R & D investments and promote economic development.* National Renewable Energy Laboratory.

National Academies of Sciences, Engineering, and Medicine. (Academy, 2016). *SBIR/STTR at the Department of Energy*. The National Academies Press. doi: https://doi.org/10.17226/23406.

National Academies of Sciences, Engineering, and Medicine. (Academy, 2020). *Review of the SBIR and STTR Programs at the Department of Energy*. The National Academies Press. doi: https://doi.org/10.17226/25674.

National Academies of Sciences, Engineering, and Medicine. (n.d.-a, Academy Assessment). *Capitalizing on science, technology, and innovation: An assessment of the small business innovation research program—Phase II.* Retrieved from https://www.nationalacademies.org/our-work/capitalizing-on-science-technology-and-innovation-an-assessment-of-the-small-business-innovation-research-program%2D%2D-phase-ii

National Academies of Sciences, Engineering, and Medicine. (n.d.-b, Academy) *Who We Are*. Retrieved from http://www.nationalacademies.org/about/whoweare/index.html

National Research Council; Committee for Capitalizing on Science, Technology, and Innovation. (2008). *An assessment of the small business innovation research program; Policy and global affairs*.

Nixon, P., Harrington, M., & Parker, D. (2012). Leadership performance is significant to project success or failure: A critical analysis. *International Journal of Productivity and Performance Management, 61*(2), 204–216.

O'Kane, C. (2016). Technology transfer executives' backwards integration: An examination of interactions between university technology transfer executives and principal investigators. *Technovation, 76-77*(2018), 64–77. https://doi.org/10.1016/j.technovation.2016.08.001

O'Kane, C., Zhang, J. A., Cunningham, J. A., & O'Reilly, P. (2017). What factors inhibit publicly funded principal investigators' commercialization activities? *Small Enterprise Research, 24*(3), 215–232.

O'Reilly, P., & Cunningham, J. A. (2017). Enablers and barriers to university technology transfer engagements with small- and medium-sized enterprises: Perspectives of principal investigators. *Small Enterprise Research, 24*(3), 274–289.

Office of the General Counsel. (2013, September). Nuclear Regulatory Legislation 112th Congress; 2nd Session. NUREG-0980, 1(10).

Ostrom, E. (1964). *Public entrepreneurship: A case study in ground water basin management'*, University of California, Los Angeles, CA, unpublished PhD dissertation. (2005). Unlocking Public Entrepreneurship and Public Economies, Working Paper DP2005/01, World Institute for Development Economic Research.

Pedraza-Farina, L. (2017). The social origins of innovation failures. *SMU Law Review, 70*, 377–446.

Pinto, J. K., & Mantel, S. J., Jr. (1990). The causes of project failure. *IEEE Transactions on Engineering Management, 37*(4), 269–276.

Rosenberg, N. (1990). Why do firms do basic research (with their own money)? *Research Policy, 19*(2), 165–174. https://doi.org/10.1016/0048-7333(90)90046-9

Rubenstein, A. H., Chakrabarti, A. K., O'Keefe, R. D., Souder, W. E., & Young, H. C. (1976). Factors influencing innovation success at the project level. *Research Management, 19*(3), 15–20.

Sauser, B. J., Reilly, R. R., & Shenhar, A. J. (2009). Why projects fail? How contingency theory can provide new insights—A comparative analysis of NASA's mars climate orbiter loss. *International Journal of Project Management, 27*(2009), 665–679.

SBIR-STTR. (2018a). *Course 1, Tutorial 2; Am I Eligible to Participate in an SBIR or STTR Program?* Retrieved from https://www.sbir.gov/sites/all/themes/sbir/dawnbreaker/img/documents/Course1-Tutorial2.pdf

SBIR-STTR. (2018b). *Birth and History of the SBIR Program*. Retrieved from https://www.sbir.gov/birth-and-history-of-the-sbir-program

SBIR-STTR. (2020). *Course 1, Tutorial 1; What is the Purpose of the SBIR and STTR Programs?* Retrieved from https://www.sbir.gov/sites/all/themes/sbir/dawnbreaker/img/documents/Course1-Tutorial1_v2.pdf

Schott, T., & Sedaghat, M. (2014). Innovation embedded in entrepreneurs' networks and national educational systems. *Small Business Economics, 43*(2), 463–476.

Shackleton, R. (2013, March). *Total factor productivity growth in historical perspective*. Working Paper Series: Congressional Budget Office.

Shepherd, D. A., & Wiklund, J. (2006). Successes and failures at research on business failure and learning from it. *Foundations and Trends in Entrepreneurship, 2*(1), 1–35.

Siegel, D. S., & Wessner, C. (2012). Universities and the success of entrepreneurial ventures: Evidence from the small business innovation research program. *Journal of Technology Transfer, 37*(4), 404–415. https://doi.org/10.1007/s10961-010-9186-3

Siegel, D. S., Wright, M., & Lockett, A. (2007). The rise of entrepreneurial activity at universities: Organizational and societal implications. *Industrial and Corporate Change, 16*(4), 489–504. https://doi.org/10.1093/icc/dtm015

Simonton, D. K. (1988). Age and outstanding achievement: What do we know after a century of research. *Psychological Bulletin, 104*(2), 251–267.

Simonton, D. K. (1991). Career landmarks in science: Individual differences and interdisciplinary contrasts. *Developmental Psychology, 27*(1), 119–130.

Singh, S., & Aggarwal, Y. (2022). In search of a consensus definition of innovation: A qualitative synthesis of 208 definitions using grounded theory approach. *Innovation: The European Journal of Social Science Research, 35*(2), 177–195. https://doi.org/10.1080/13511610.2021.1925526

Spiller, P. T., & Teubal, M. (1993). Analysis of R&D failure. *Research Policy, 22*(2), 113–113.

The Associated Press. (1993, August 15). 14-Year Cleanup at Three Mile Island Concludes. *The New York Times*. Retrieved from https://www.nytimes.com/1993/08/15/us/14-year-cleanup-at-three-mile-island-concludes.html

Thursby, J. G., Jensen, R., & Thursby, M. C. (2001). Objectives, characteristics and outcomes of university licensing: A survey of major U. S. universities. *The Journal of Technology Transfer, 26*(1/2), 59–72. https://doi.org/10.1023/A:1007884111883

Toole, A. A., & Czarnitzki, D. (2009). Exploring the relationship between scientist human capital and firm performance: The case of biomedical academic entrepreneurs in the SBIR program. *Management Science, 55*(1), 101–114. https://doi.org/10.1287/mnsc.1080.0913

Wagner, R. E. (1966). Pressure groups and political entrepreneurs: A review article. *Public Choice, 1*, 161–170.

White House—Office of the Press Secretary. (1987). *The President's superconductivity initiative; President Announces 11-Point Superconductivity initiative*. DOE This Month.

Index

M. Boyce, *Principal Investigators and R&D Failure*, Academic Entrepreneurship, Scientists, and Scientific Careers, https://doi.org/10.1007/978-3-031-43608-6

OK

Advances in Intelligent Systems and Computing

Volume 1347

The series "Advances in Intelligent Systems and Computing" contains publications on theory, applications, and design methods of Intelligent Systems and Intelligent Computing. Virtually all disciplines such as engineering, natural sciences, computer and information science, ICT, economics, business, e-commerce, environment, healthcare, life science are covered. The list of topics spans all the areas of modern intelligent systems and computing such as: computational intelligence, soft computing including neural networks, fuzzy systems, evolutionary computing and the fusion of these paradigms, social intelligence, ambient intelligence, computational neuroscience, artificial life, virtual worlds and society, cognitive science and systems, Perception and Vision, DNA and immune based systems, self-organizing and adaptive systems, e-Learning and teaching, human-centered and human-centric computing, recommender systems, intelligent control, robotics and mechatronics including human-machine teaming, knowledge-based paradigms, learning paradigms, machine ethics, intelligent data analysis, knowledge management, intelligent agents, intelligent decision making and support, intelligent network security, trust management, interactive entertainment, Web intelligence and multimedia.

The publications within "Advances in Intelligent Systems and Computing" are primarily proceedings of important conferences, symposia and congresses. They cover significant recent developments in the field, both of a foundational and applicable character. An important characteristic feature of the series is the short publication time and world-wide distribution. This permits a rapid and broad dissemination of research results.

Indexed by DBLP, EI Compendex, INSPEC, WTI Frankfurt eG, zbMATH, Japanese Science and Technology Agency (JST).

All books published in the series are submitted for consideration in Web of Science.

More information about this series at http://www.springer.com/series/11156

Sajid Anwar · Abdul Rauf
Editors

Proceedings of the First International Workshop on Intelligent Software Automation

ISEA 2020

Springer

Editors
Sajid Anwar
Center of Excellence in IT
Institute of Management
Sciences Peshawar
Peshawar, Pakistan

Abdul Rauf
Research Institute
of Sweden in Vasteras
Vasteras, Sweden

ISSN 2194-5357 ISSN 2194-5365 (electronic)
Advances in Intelligent Systems and Computing
ISBN 978-981-16-1044-8 ISBN 978-981-16-1045-5 (eBook)
https://doi.org/10.1007/978-981-16-1045-5

This Springer imprint is published by the registered company Springer Nature Singapore Pte Ltd.
The registered company address is: 152 Beach Road, #21-01/04 Gateway East, Singapore 189721, Singapore

Preface

The need for novel software engineering (SE) tools and techniques which are highly reliable and greatly robust are the order of the day. There is a greater understanding that the design and evolution of the software systems and tools must be 'smart', if it is to remain efficient and effective. The nature of artifacts, from specifications to delivery, produced during construction of software systems, can be very convoluted and difficult to manage. A software engineer cannot find all its intricacies by examining these artifacts manually. Automated tools and techniques are required to reflect over business knowledge to identify what is missing or could be effectively changed while producing and evolving these artifacts. There is an agreed belief among researchers that SE provides an ideal platform to apply and test the recent advances in AI (Artificial Intelligence) tools and techniques. More and more SE problems are now resolved through the application of AI, such as through tool automation.

The international workshop on Intelligent Software Engineering Automation (ISEA) was an initiative to consider the above-mentioned considerations and challenges. Intelligent Software Engineering Automation (ISEA) aimed at bringing together international researchers and practitioners in the fields of intelligent Software Engineering/Automated software engineering to present and discuss applications, experiences and emerging advanced techniques. To this end, the workshop welcomed original articles on every aspect related to intelligent software automation. Workshop aimed at contributing to having scientific novelty as well as fostering discussion and networking by:

1. highlighting the areas in SE, which can benefit from intelligent techniques and its optimization,
2. discovering the potential benefits and possible challenges of using intelligent techniques in software engineering automation activities, and
3. developing a community of researchers to cater for issues of intelligence in SE.

Among attendees of our workshop were scientists and representatives of IT companies as well as researchers and academicians working in the field of Intelligent Software Engineering Automation. ISEA was the workshop organized under the technical co-sponsorship of the Asia-Pacific Software Engineering Conference (APSEC2020). We also continued our successful cooperation with Springer, which resulted in the publication of this book. ISEA had fifteen participants from three countries, which made this workshop a successful and memorable event.

ISEA 2020 workshop is focused on all aspects of intelligent software automation, and this book consists of research papers presented in the workshop related to the following topics:

1. Scalable AI algorithms for mining large scale software repositories
2. Machine learning techniques for software maintenance prediction
3. Deep learning models for software development and production challenges
4. Software architecture knowledge extraction using ML
5. Software optimization using evolutionary algorithms
6. Mining software specifications
7. AI-based computer-assisted coding
8. Code Churn Prediction
9. Technical debt identification using ML
10. Intelligent Software Summarization
11. Program comprehension using ML
12. Intelligent software traceability
13. AI models/techniques for software artifacts' quality evaluation
14. Advancement in intelligent technologies for SE
15. Intelligent software reverse engineering

We would like to thank all Program Committee members, as well as the additional reviewers, for their effort in reviewing the papers. We would like to extend special thanks to Mr. Adnan Ameen (lecturer, Institute of Management sciences Peshawar, Pakistan)—builder and administrator of the workshop website http://isea2020.b-softs.com.

We hope that the topics covered in Intelligent Software Engineering Automation workshop proceedings will help the readers understand the intricacies involving the methods and tools of software engineering that have become an important element of nearly every branch of computer science.

Peshawar, Pakistan
Vasteras, Sweden

Sajid Anwar, Ph.D.
Abdul Rauf, Ph.D.

Contents

Editors and Contributors

About the Editors

Dr. Sajid Anwar is Associate Professor at the Center of Excellence in Information Technology Institute of Management Sciences (IMSciences), Peshawar, Pakistan. He received his M.S. (Computer Science, 2007) and Ph.D. degrees (Software Engineering, 2011) from NUCES-FAST, Islamabad. Previously, he was Head of Undergraduate Program in Software Engineering at IMSciences. Dr. Sajid Anwar is Leading Expert in software architecture engineering and software maintenance prediction. His research interests are cross-disciplinary and industry focused and includes search-based software engineering; Prudent based expert systems; customer analytics; active learning and applying data mining and machine learning techniques to solve real-world problems. Dr. Sajid Anwar is Associate Editor of Expert Systems Journal Wiley. He has been Guest Editor of numerous journals, such as Neural Computing and Applications, Cluster Computing Journal Springer, Grid Computing Journal Springer, Expert Systems Journal Wiley, Transactions on Emerging Telecommunications Technologies Wiley, and Computational and Mathematical Organization Theory Journal Springer. He is also in Member Board Committee Institute of Creative Advanced Technologies, Science and Engineering, Korea (iCatse.org). He has supervised to completion of many M.S. research students. He has conducted and led collaborative research with government organizations and academia and has published over 45 research articles in prestigious conferences and journals.

Dr. Abdul Rauf is currently working as ERCIM Postdoc Fellow at RISE Research Institutes of Sweden in Vasteras, Sweden. He received his Ph.D. (2011) from National University of Computer and Emerging Sciences, Pakistan. He has also worked as Assistant Professor in Imam Saud University in Saudi Arabia from 2012 to 2019. He has published 40+ international peer-reviewed conference and journal articles. His research interests relate to software testing, natural language processing, and social media analysis. Other areas of interest include software process improvement, quality assurance, software requirement engineering, and software project management. He has been in organizing committee of the ICTHP and was Track Chair at ICACC. He was also part of the technical committee in several conferences including FIT, INMIC, ICET, ICEET, IMDC, DSDE, ECIST, and IntelliSys.

Contributors

Syed Manzar Abbas Department of Computer Science, FAST-NUCES, Islamabad, Pakistan

Mohsin Ahmed Department of Computer Science, COMSATS University Islamabad, Islamabad, Pakistan

Khubaib Amjad Alam Department of Computer Science, National University of Computer and Emerging Sciences (FAST-NUCES), Islamabad, Pakistan

Rimsha Asif Department of Computer Science, National University of Computer and Emerging Sciences (FAST-NUCES), Islamabad, Pakistan

Tulshi Chandra Das IIT, University of Dhaka, Dhaka, Bangladesh

Abdullah Al Jubaer IIT, University of Dhaka, Dhaka, Bangladesh

Saif Ur Rehman Khan Department of Computer Science, COMSATS University Islamabad, Islamabad, Pakistan

Kwang Man Ko Department of Computer Engineering, Sangji University, Wonju-si, Republic of South Korea

Bisma Rehman Department of Computer Science, National University of Computer and Emerging Sciences (FAST-NUCES), Islamabad, Pakistan

Maloy Kanti Sarker IIT, University of Dhaka, Dhaka, Bangladesh

Md. Shihab Shohrawardi IIT, University of Dhaka, Dhaka, Bangladesh

Md Saeed Siddik IIT, University of Dhaka, Dhaka, Bangladesh

A Three-Way Decision-Making Approach for Customer Churn Prediction Using Game-Theoretic Rough Sets

Syed Manzar Abbas, Khubaib Amjad Alam, and Kwang Man Ko

Abstract Churn prediction models play a crucial role in improving Customer relationship management (CRM). However, most of the existing Churn prediction models utilize classification schemes during analytical computation which hinders concrete decision-making and concern of misclassification of customers due to the lack of essential information is always there. In this article, we address the problem of misclassification in the churn prediction model due to data sparsity. For this purpose, a Three-way classification (TWC) approach using Game-theoretic Rough sets (GTRS) has been used in this research. The purpose of using TWC is to make a third sensible decision for the classification of objects with missing information, thereby reducing the probability of misclassification. A major issue in TWC is the selection of an appropriate threshold to control the three-way decisions. Hence, we have used GTRS for computing effective thresholds by determining the tradeoff between the effectiveness and coverage of the classification. The proposed approach is evaluated on cell2cell and IBM Telecom churn datasets, and experimental results signify that the use of TWC can significantly improve the overall churn classification accuracy.

Keywords Three-way classification · Churn prediction · Game theory · Rough sets · Predictive analytics

S. M. Abbas (✉) · K. A. Alam
Department of Computer Science, FAST-NUCES, Islamabad, Pakistan
e-mail: f179002@nu.edu.pk

K. A. Alam
e-mail: khubaib.amjad@nu.edu.pk

K. M. Ko
Department of Computer Engineering, Sangji University, Wonju-si, Republic of South Korea
e-mail: kkman@sangji.ac.kr

S. Anwar and A. Rauf (eds.), *Proceedings of the First International Workshop on Intelligent Software Automation*, Advances in Intelligent Systems and Computing 1347, https://doi.org/10.1007/978-981-16-1045-5_1

1 Introduction

Customer churn prediction is a fundamental mechanism in predictive analytics which aims to devise strategic approaches towards customer retention [1]. Generally, the classification approaches are considered as the most suitable approach for devising churn prediction scheme [2]. Hence, a churn prediction model (CPM) is often considered as a binary classification problem in which the customers are classified into two groups through definite decision [1]. However, using definite decision-making through binary classification in CPM reduces the overall accuracy of the classification scheme in case of flaky conditions which commonly arise due to incomplete historic information of customers [3]. Therefore, binary classification with high effectiveness may have limited coverage and classification with high coverage may have lesser effectiveness [2]. It should be noted here that the effectiveness of classification reflects the ratio of correctly classified items by the total number of items. Likewise, the term coverage expresses the proportion of classified items from the total number of items. Hence, increasing the coverage rate for classifying the legitimate customers as churners (detractors) will also increase the probability rate of classifying customers with flaky details as non-churners (loyal). Therefore, such a classification scheme may reduce the accuracy of CPM by misclassifying the legitimate customer with flaky information into the non-churner group [4]. Likewise, misclassifying a loyal customer to churner also causes extra cost to the business due to the waste of resources on retaining such customers [3].

The major contribution of this study is the improvement of classification approach by optimizing the effectiveness and coverage in CPM. To this purpose, we proposed a three-way classification scheme using Game-theoretic rough sets (TWC-GTRS) [5]. The main advantage of using three-way classification (TWC) is that it makes three-way decision which includes accepting an item for classifying in a group, rejection from the membership of a group, and deferring from a definite decision of accepting or rejecting an item as belonging to a group [6]. This approach has been a practice in cases where data sparsity exists in the provided information. As interesting as it may appear, TWC critically depends upon the configuration of thresholds which control the type of decisions [5]. To this end, we have used Game-theoretic Rough Sets (GTRS) for obtaining effective thresholds configuration by forming a game to determine a tradeoff between effectiveness and coverage of classification in CPM.

The forthcoming sections are organized as follows: Section 2 presents the related work. The proposed methodology is defined in Sect. 3. Likewise, the experimental results are discussed in Sect. 4. Lastly, the conclusion is given in Sect. 5.

2 Related Work

The term churn prediction refers to the mechanism for advance identification of the customers who are likely to switch from one brand to another competitor brand. In general, the churners are classified as those customers who switch from one brand to a competitor's brand [1]. According to the literature, the resource cost for adding a new customer to the business is higher than retaining the existing customers. Therefore, the efficacious prediction of customer reviews/ratings a vital role to improve the business [7, 8]. The most common approach for developing CPM is by using decision trees which uses binary classification for intelligently classifying the customers [3, 4]. However, the major limitation in this study is the lack of optimizing the classification coverage rate. Relatedly, k-nearest neighbors (kNN) and Naïve bayes approach has also been used for devising the CPMs [3]. But the major drawback through these approaches on large datasets is a decrease in accuracy because of definite decisions using binary classification. Besides the typical classification approaches, we can cite some state-of-the-art advance approaches such as Artificial Neural network (ANN) and rough set (RST) based approaches which use a binary scheme for the training of the model [1, 4]. However, the performance of such approaches is weak due to the misclassification of the customers with flaky details. In particular, the misclassification in CPM is the result of high expectation for better effectiveness and coverage in classification [1].

From the literature, it has been identified that despite some differences in the typically used approaches for developing CPMs, the core focus is on utilizing some sort of binary classification scheme by categorizing customers either as churners or not. Generally speaking, a binary classification faced with the issue of misclassification of items in case of high expectation for coverage and effectiveness [7]. The prior attempts for improving CPMs motivates that adopting an advanced binary classification algorithm is not a proper solution. There is a need of introducing a better scheme which can handle the misclassification in CPM. To address this issue, we extended the application of three-way classification using game-theoretic rough sets (TWC-GTRS) for developing a novel model to improve the effectiveness of CPMs. In particular, the main contribution of our research work is different from extant researches in two ways. Firstly, our work addresses the generic problem of binary classification in CPMs. Secondly, we focus upon the two major properties of classification (effectiveness and coverage) simultaneously in order to obtain effective predictions. Moreover, our work applies TWC-GTRS in CPM. We have used TWC to introduce a new category of deferment from classification for improving effectiveness and coverage. In addition, GTRS have been used to compute filters, for defining the categorizing of an item in deferred class and for categorizing the item of deferred class into churners or non-churners based upon further collected information.

3 TWC-GTRS Based Churn Prediction

This section elaborates the concepts of TWC-GTRS based CPM along with the working methodology of our proposed scheme.

3.1 Rudimentary Concepts of Three-Way Classification

In general, a three-way classification (TWC) approach categorizes group items into three pairwise disjoint regions [9]. The essential idea behind this approach is to make a deferment rejoin for categorizing the items with insufficient information whenever it is not clear and possible to decide whether or not to include an object in a group. Generally speaking, a universal set U of items with equivalence class X can be divided into three regions namely, acceptance *POS(X)*, rejection *NEG(X)*, and deferment region *BND(X)* [10]. TWC uses probabilistic thresholds for taking the decision to classify an item into one of these three regions. In particular, Pawlak rough sets define TWC through conditional probability $\mathcal{P}(|[y])$ which demonstrate the probability of an object y in equivalence class X [11]. The three-way regions of X with thresholds (α, β) can be defined as:

$$POS_{(\alpha,\beta)}(\mathcal{X}) = \{y \in U \mid \mathcal{P}(\mathcal{X}|[y]) \geqslant \alpha\}, \tag{1}$$

$$NEG_{(\alpha,\beta)}(\mathcal{X}) = \{y \in U \mid \mathcal{P}(\mathcal{X}|[y]) \leqslant \beta\}, \tag{2}$$

$$BND_{(\alpha,\beta)}(\mathcal{X}) = \{y \in U \mid \beta < \mathcal{P}(\mathcal{X}|[y]) < \alpha\}. \tag{3}$$

Where an object y will be accepted for classification in class X if the conditional probability is greater than or equal to acceptance threshold α, i.e. $\mathcal{P}(\mathcal{X}|[y]) \geqslant \alpha$. In contrast, object y will be rejected from X if the conditional probability is less than or equal to rejection threshold β, i.e., $\mathcal{P}(\mathcal{X}|[y]) \leqslant \beta$. In situations, where sufficient information of object y cannot be determined such that conditional probability is between the thresholds α and β, i.e., $\beta < \mathcal{P}(\mathcal{X}|[y]) < \alpha$, the decision for classifying or rejecting y from X will be deferred. However, it has been identified from the literature that using TWC for obtaining an effective configuration of threshold is practically not possible [5].

3.2 Rudimentary Concepts of Game-Theoretic Rough Sets

The essential idea behind game-theoretic rough sets (GTRS) is to formulate a game to reach a tradeoff between involved criteria, i.e., probabilistic thresholds [5]. GTRS is divided into two phases which are defined as
Game Formulation: In general, a game G can be defined as a tuple $\{P, A, \mu\}$ containing three basic elements:

- P represents the set of players, $P = \{p_1, p_2, p_3, \ldots, p_n\}$.
- A depicts the set of actions, $A = A_1 \times A_2 \times \cdots \times A_n$, where each A_i denotes the strategy set for player p_i.
- μ denotes the set of pay-off function, where $\mu = (\mu_1, \mu_2, \mu_3, \ldots, \mu_n)$, where μ_i: $A \mapsto \mathbb{R}$ is a real-valued pay-off function for player p_i.

In particular, GTRS can be used in TWC for computing effective thresholds by configuring tradeoff between effectiveness and coverage of classification [6]. To that purpose, a game can be formulated with players as effectiveness $\mathcal{E}$ and coverage $\mathcal{C}$, such that each player p_i of the game G selects an optimal action A_i for maximizing their own pay-off μ_i.

Learning Approach: In general, Nash equilibrium is typically adopted for computing the possible game outcomes in GTRS [5, 6]. This equilibrium represents a compromise between the involved criteria players based upon best action choices. Hence, with a basic goal of the game to maximize both the effectiveness and coverage of classification, this equilibrium can be used to depict the tradeoff between the involved players, i.e., $\mathcal{E}$ and $\mathcal{C}$. In addition, this iteration stops when the overall pay-off gain μ_i does not improve through new actions.

3.3 Three-Way Classification for Churn Prediction

In recent years, a variety of churn prediction models have emerged, however, there are still some problems such as sparsity that are not well solved. This sparsity issue commonly leads to the lack of sufficient information for classifying an item into churner or non-churner groups accurately. To boost the prediction accuracy, a three-way classification can efficaciously address the problem of sparsity in data.

A simple approach for obtaining a three-way classification of churners can be perform using Eqs. (1)–(3). The major focus in this process is to measure the two important properties of classification which are effectiveness and coverage. It has been identified that the performance of a classification process can be evaluated through these two properties. Moreover, the effectiveness of classification reflects the ratio of correctly classified items by a total number of items as shown in the equation below, where CU represents the churners and CU^c depicts the complement of churners.

$$\mathcal{E}(CU) = \frac{|CU \cap POS_{(\alpha,\beta)}(CU)| + |CU^c \cap NEG_{(\alpha,\beta)}(CU)|}{|POS_{(\alpha,\beta)}(CU)| + |NEG_{(\alpha,\beta)}(CU)|} \tag{4}$$

Likewise, the term coverage expresses the proportion of classified items from a total number of items as shown in the equation below.

$$\mathcal{C}(CU) = \frac{|POS_{(\alpha,\beta)}(CU)| + |NEG_{(\alpha,\beta)}(CU)|}{|U|} \tag{5}$$

3.4 GTRS for Churn Prediction

The selection of effective threshold configuration is important for the accuracy of TWC [11]. To this end, it has been identified from extant researches that GTRS can be use for computing thresholds to be used in TWC [11]. For the sake of clarity, the involved steps are defined below

Game Formulation In general, a game G contains three basic elements, namely, set of players P, set of actions A, and the set of pay-off function μ. Hence, G can be represented as a tuple $\{P, A, \mu\}$.

Players: In our case, the possible game players can be the effectiveness and coverage $P = \{\mathcal{E}, \mathcal{C}\}$ of classification.

Actions: The possible set of actions for both $\mathcal{E}$ and $\mathcal{C}$ can be denoted as $A = \mathrm{A}_{\mathcal{E}} \times \mathrm{A}_{\mathcal{C}}$, where $\mathrm{A}_{\mathcal{E}} = \{\mathrm{a}_{\mathcal{E}1}, \mathrm{a}_{\mathcal{E}2}, \cdots, \mathrm{a}_{\mathcal{E}k}\}$ represents the set of actions for player $\mathcal{E}$ and $\mathrm{A}_{\mathcal{C}} = \{\mathrm{a}_{\mathcal{C}1}, \mathrm{a}_{\mathcal{C}2}, \cdots, \mathrm{a}_{\mathcal{C}k}\}$ depicts the available actions for the player $\mathcal{C}$. These actions are the major responsive factors for changing the given thresholds. Moreover, the possible actions that can be performed by the given player $\mathcal{E}$ can be decreasing of β such as, $\mathrm{A}_{\mathcal{E}} = \{\text{Decreasing } \beta, \text{ No change on } \beta\}$. Likewise, the possible actions that can be performed by the given player $\mathcal{C}$ can be increasing of α such as, $\mathrm{A}_{\mathcal{C}} = \{\text{Increasing } \alpha, \text{ No change on } \alpha\}$.

Pay-off Functions: The pay-off for each of the player can be denoted as $\mu = \{\mu_{\mathcal{E}}, \mu_{\mathcal{C}}\}$. In general, if the player $\mathcal{E}$ performing an action $\mathrm{a}_{\mathcal{E}}$ and the player $\mathcal{C}$ performing an action $\mathrm{a}_{\mathcal{C}}$, the pay-off function for player $\mathcal{E}$ and $\mathcal{C}$ will become $\mu_{\mathcal{E}}(\mathrm{a}_{\mathcal{E}}, \mathrm{a}_{\mathcal{C}})$ and $\mu_{\mathcal{C}}(\mathrm{a}_{\mathcal{E}}, \mathrm{a}_{\mathcal{C}})$ respectively. The actions $\mathrm{a}_{\mathcal{E}}$ and $\mathrm{a}_{\mathcal{C}}$ performed by the players $\mathcal{E}$ and $\mathcal{C}$ actually reflects the changes for the thresholds $\{\alpha, \beta\}$. Hence, it can be concluded that these thresholds are actually the result of changes by performing the actions $\mathrm{a}_{\mathcal{E}}$ and $\mathrm{a}_{\mathcal{C}}$ by the players of the game. In particular, the pay-off functions $\mu_{\mathcal{E}}(\alpha, \beta)$ and $\mu_{\mathcal{C}}(\alpha, \beta)$ for the changes in thresholds through actions can be defined using pay-off table.

Learning Approach In general, Nash equilibrium is typically adopted for computing the possible game outcomes in GTRS [5]. This algorithm repeatedly iterates for identifying that either there are any other better configuration available than the current choice obtained through Nash equilibrium. In addition, the stopping criteria of the game can be defined in a way if the current thresholds (α^1, β^1) are not good enough for the configuration then another iteration will take place and produce some new (α^2, β^2). If the overall pay-off gain μ_i does not improve through new actions then the system will stop the iterations. The proposed algorithm is given below

Algorithm 1: TWC-GTRS for Churn Classification

```
Input: 𝒫 = Players, a = Strategies
Output: Three-way classification of churners
Initialize α = 1.0 and β = 0.0
Compute ℰ and 𝒞 using Eqs. (4) and (5)
for each a_n ={Strategies} from a; do
Calculate pay-off functions μ_ℰ(α, β) and μ_𝒞(α, β).
Populate the pay-off table
Calculate Nash equilibrium for μ_ℰ and μ_𝒞
if (μ_ℰ(α, β) ⩽ μ_𝒞(α, β)) || ((α^1, β^1) ⩾ (α*, β*))
Record current threshold configuration (α, β)
else
Use other set of actions from the pay-off table
end if
Apply TWC using computed configuration of (α, β).
end for
```

4 Experimental Results and Discussion

4.1 Data Preprocessing

For the sake of evaluating the proposed approach, we have used cell2cell [2] and IBM telecom datasets [4]. The dataset of cell2cell have 51,000 entries where 71% of this dataset is marked as non-churners and 29% data is marked as churners [1]. Similarly, the IBM telecom dataset have around 70,000 entries where 75% entries are non-churners and 25% of the entries are churner. Moreover, we have use holdout validation strategy to split both the datasets into 80–20%, where 80% of the data has been used for the training of the model and the remaining 20% has been used for the evaluation of the model. The justification of considering holdout approach rather than the cross-validation is that the size of the selected datasets is large and it has been observed that low variance is reported on splitting the dataset into 80–20 sections. Hence, to avoid the extra computational cost, we have use holdout validation strategy.

4.2 Threshold Configuration Computing

Based upon the proposed algorithm, we have formulated a game between effectiveness $\mathcal{E}$ and coverage $\mathcal{C}$ of the classification of the churners. The initial threshold configuration of the game is $(1, 0.5)$. In particular, the goal of this game is to establish a tradeoff between the effectiveness and coverage of the classification performance. To this purpose, the player $\mathcal{E}$ tries to decrease the β with a step change of 0.02, 0.04, 0.1, and 0.4 in both datasets. Likewise, the player $\mathcal{C}$ tries to decrease the α with a step change of 0.01, 0.03, 0.08, and 0.1. The results for the pay-off computation for the both datasets are reported in Tables 1, 2, and 3. We set the stopping criteria in our experiment as the value of $\mathcal{E}$ is greater than the predefined value 99. The resultants we obtained for both datasets of cell2cell and IBM are $< 0.9908, 0.9189 >$ and $< 0.9879, 0.9017 >$, respectively.

4.3 Repeatition Learning

Based upon the pay-off tables, it can be observed that we have repeated the game for four times in both cases as shown in Tables 2 and 4. The purpose of these repetitions is to identify the effectiveness value greater than 0.99 and the appropriate threshold configuration. Hence, on the fourth step, we have identified that the most suitable threshold tradeoff for both datasets of IBM telecom and cell2cell is $(\alpha, \beta) = (0.78, 0.08)$.

Table 1 Pay-off values for "*cell2cell*" dataset using *TWC-GTRS*

		$\mathcal{C}$				
		α	$\alpha\downarrow 0.02$	$\alpha\downarrow 0.04$	$\alpha\downarrow 0.1$	$\alpha\downarrow 0.4$
$\mathcal{E}$	β	$< 0.9519, 0.9327 >$	$< 0.9519, 0.9327 >$	$< 0.9519, 0.9327 >$	$< 0.9519, 0.9327 >$	$< 0.9519, 0.9327 >$
	$\beta\downarrow 0.01$	$< 0.9520, 0.9315 >$	$< 0.9520, 0.9315 >$	$< 0.9520, 0.9315 >$	$< 0.9520, 0.9315 >$	$< 0.9520, 0.9315 >$
	$\beta\downarrow 0.03$	$< 0.9684, 0.9297 >$	$< 0.9684, 0.9297 >$	$< 0.9684, 0.9297 >$	$< 0.9684, 0.9297 >$	$< 0.9684, 0.9297 >$
	$\beta\downarrow 0.08$	$< 0.9874, 0.9205 >$	$< 0.9874, 0.9205 >$	$< 0.9874, 0.9205 >$	$< 0.9874, 0.9205 >$	$< 0.9874, 0.9205 >$
	$\beta\downarrow 0.1$	$< 0.9908, 0.9189 >$	$< 0.9908, 0.9189 >$	$< 0.9908, 0.9189 >$	$< 0.9908, 0.9189 >$	$< \mathbf{0.9908, 0.9189} >$

Table 2 Repeatition learning for "*cell2cell*" dataset

Initial configuration (α, β)	Actions	Resultant (α, β)	Pay-off results
$(1, 0.5)$	$(\mathcal{E}\downarrow 0.01, \mathcal{C}\downarrow 0.02)$	$(0.99, 0.48)$	$< 0.9908, 0.9189 >$
$(0.99, 0.48)$	$(\mathcal{E}\downarrow 0.03, \mathcal{C}\downarrow 0.04)$	$(0.96, 0.44)$	$< 0.9874, 0.9205 >$
$(0.96, 0.44)$	$(\mathcal{E}\downarrow 0.08, \mathcal{C}\downarrow 0.1)$	$(0.88, 0.34)$	$< 0.9684, 0.9297 >$
$(0.88, 0.34)$	$(\mathcal{E}\downarrow 0.1, \mathcal{C}\downarrow 0.5)$	$(0.78, 0.08)$	$< 0.9520, 0.9315 >$

Table 3 Pay-off values for "*IBM Telecom*" dataset using *TWC-GTRS*

		$\mathcal{C}$				
		α	$\alpha \downarrow 0.02$	$\alpha \downarrow 0.04$	$\alpha \downarrow 0.1$	$\alpha \downarrow 0.5$
$\mathcal{E}$	β	$< 0.9406, 0.9287 >$	$< 0.9406, 0.9287 >$	$< 0.9406, 0.9287 >$	$< 0.9406, 0.9287 >$	$< 0.9406, 0.9287 >$
	$\beta \downarrow 0.01$	$< 0.9497, 0.9374 >$	$< 0.9497, 0.9374 >$	$< 0.9497, 0.9374 >$	$< 0.9497, 0.9374 >$	$< 0.9497, 0.9374 >$
	$\beta \downarrow 0.03$	$< 0.9575, 0.9239 >$	$< 0.9575, 0.9239 >$	$< 0.9575, 0.9239 >$	$< 0.9575, 0.9239 >$	$< 0.9575, 0.9239 >$
	$\beta \downarrow 0.08$	$< 0.9798, 0.9163 >$	$< 0.9798, 0.9163 >$	$< 0.9798, 0.9163 >$	$< 0.9798, 0.9163 >$	$< 0.9798, 0.9163 >$
	$\beta \downarrow 0.1$	$< 0.9879, 0.9017 >$	$< 0.9879, 0.9017 >$	$< 0.9879, 0.9017 >$	$< 0.9879, 0.9017 >$	$< \mathbf{0.9901}, \mathbf{0.9017} >$

Table 4 Repeatition learning for "*IBM Telecom*" dataset

Initial configuration (α, β)	Actions	Resultant (α, β)	Pay-off results
(1, 0.5)	$(\mathcal{E}\downarrow 0.01, \mathcal{C}\downarrow 0.02)$	(0.99, 0.48)	$< 0.9901, 0.9017 >$
(0.99, 0.48)	$(\mathcal{E}\downarrow 0.03, \mathcal{C}\downarrow 0.04)$	(0.96, 0.44)	$< 0.9798, 0.9163 >$
(0.96, 0.44)	$(\mathcal{E}\downarrow 0.08, \mathcal{C}\downarrow 0.1)$	(0.88, 0.34)	$< 0.9575, 0.9239 >$
(0.88, 0.34)	$(\mathcal{E}\downarrow 0.1, \mathcal{C}\downarrow 0.5)$	(0.78, 0.08)	$< 0.9497, 0.9374 >$

4.4 Evaluation Measure

For the evaluation of the proposed model, the observed measures are observing the performance of the model in terms of classifying the maximum items accurately. However, the tradeoff between the accuracy and coverage has also been considered for the evaluation purpose.

Effectiveness and Coverage We have computed the accuracy of the proposed model by comparing it with state-of-the-art baseline researches, namely random forest (RF) based approach for classification proposed by [2], neural network (ANN) based classification of churners proposed in [1], and rough set (RST) based classification of churners proposed by authors in [4].

For the sake of computing the accuracy of the model, we have used recall@n, where n is the number of observations. The performance results obtained using recall are depicted in Figs. 1 and 2 for both datasets. Moreover, the summary of the results

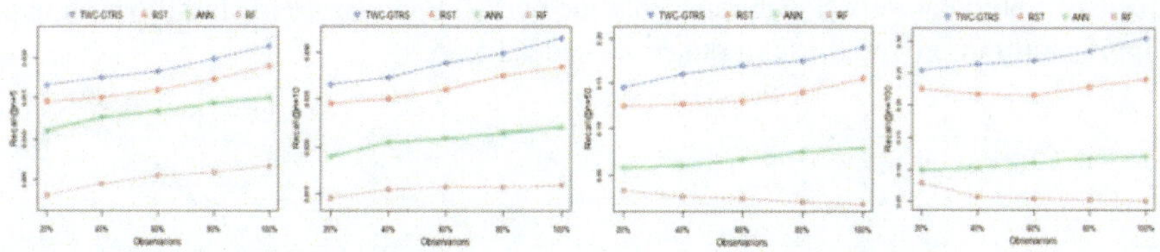

Fig. 1 Reported results of recall@n with n = 5, 10, 50, 100 for *cell2cell* dataset for *TWC-GTRS*

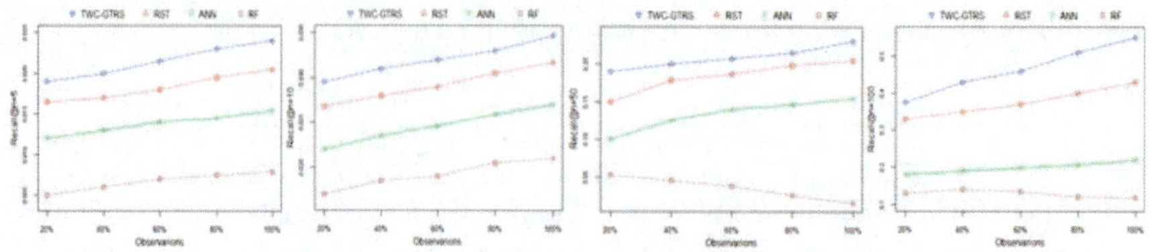

Fig. 2 Reported results of recall@n with n = 5, 10, 50, 100 for *IBM Telecom* for *TWC-GTRS*

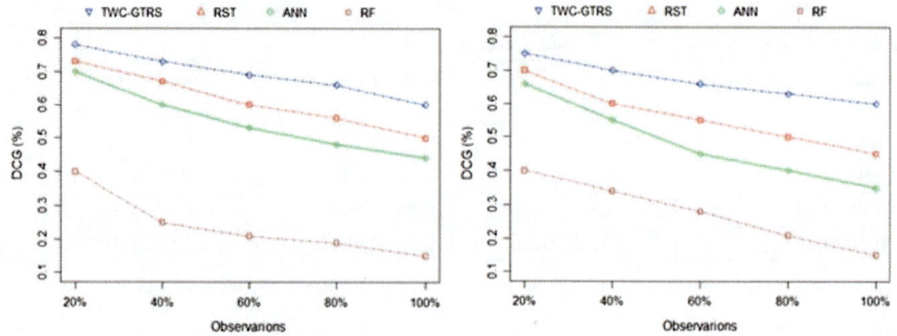

(a) Results for DCG on cell2cell dataset

(b) Results for DCG on IBM Telecom dataset

Fig. 3 Results of DCG on *cell2cell* and *IBM Telecom* dataset

is reported in Table 5. In particular, an improvement of 9% can be observed by the blue line in both cases of datasets. Interestingly, the recall values of our approach for both datasets are higher in case of changing the number of observations. Hence, it can be concluded that the performance of the proposed model is comparatively better. Likewise, for the sake of computing the performance of the coverage of the classification using our proposed scheme TWC-GTRS, we have used distributed cumulative gain (DCG). The results of this evaluation are depicted in Fig. 3a, b for both datasets cell2cell and IBM Telecom, respectively.

Performance of TWC-GTRS In order to validate the performance of TWC-GTRS, we have used the error computing approaches RMSE and MAE. The results of both approaches are visualized in Fig. 4. The result signifies that the lowest error for both approaches RMSE and MAE has been reported through our proposed methodology in case of both datasets. This means that the overall uncertainty in classification has significantly improved through our proposed scheme.

Table 5 Reported results of recall for *cell2cell* and *IBM Telecom* dataset using *TWC-GTRS*

Approaches	Dataset of cell2cell				Dataset of IBM Telecom			
	Recall@n				**Recall@n**			
	n = 5	n = 10	n = 50	n = 100	n = 5	n = 10	n = 50	n = 100
RF	+3.04%(± 0.01)	+ 4.59%(± 0.01)	+5.79%(± 0.01)	+7.04%(± 0.01)	+ 6.83%(± 0.01)	+ 3.91%(± 0.01)	+ 3.58%(± 0.01)	+ 1.09%(± 0.01)
ANN	+11.07%(± 0.01)	+17.81%(± 0.01)	+19.63%(± 0.01)	+20.09%(± 0.01)	+14.07%(± 0.01)	+ 12.97%(± 0.01)	+ 9.96%(± 0.01)	+ 5.70%(± 0.01)
RST	+14.19%(± 0.01)	+ 24.87%(± 0.01)	+ 26.99%(± 0.01)	+ 28.06%(± 0.01)	+14.92%(± 0.01)	+12.72%(± 0.01)	+ 9.91%(± 0.01)	+ 5.81%(± 0.01)
TWC-GTRS	**+26.79%(± 0.01)**	**+28.31%(± 0.01)**	**+33.82%(± 0.01)**	**+35.70%(± 0.01)**	**+23.16%(± 0.01)**	**+18.24%(± 0.01)**	**+15.11%(± 0.01)**	**+12.71%(± 0.01)**

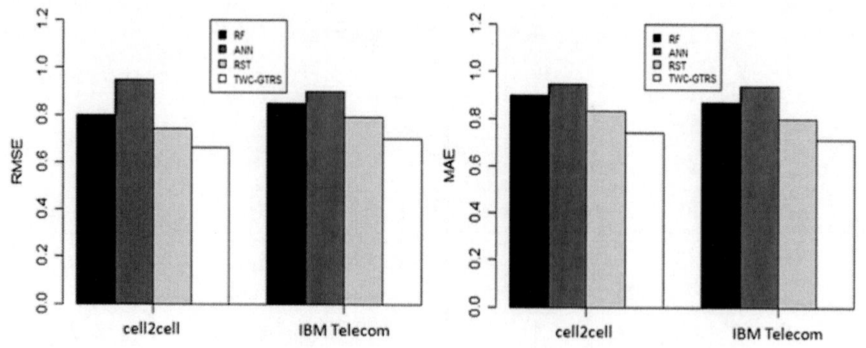

Fig. 4 Results of RMSE and MAE on *cell2cell* and *IBM Telecom* dataset

5 Conclusion and Future Directions

A binary classification approach in churn prediction models is faced with two inherited limitations of definite decision-making and the concern of misclassification of customers due to lack of essential information. In this article, we examine the role of three-way classification for improving the two major properties effectiveness and coverage of the traditional churn classification model using three-way classification approach based on game-theoretic rough sets (TWC-GTRS). The main purpose of using GTRS is to select an effective threshold configuration for the TWC to perform an accurate classification. Experimental results on the datasets of IBM telecom and cell2cell suggest that the use of TWC can improve the overall accuracy churn classification up to 9% in comparison to the of traditional baseline approaches.

Acknowledgements This research was supported by Basic Science Research Program through the National Research Foundation of Korea (NRF) funded by the Ministry of Education (2017030223).

References

1. Mishra, Abinash and U. Srinivasulu Reddy. 2017. A novel approach for churn prediction using deep learning. In *2017 IEEE international conference on computational intelligence and computing research (ICCIC)*, 1–4. Piscataway, NJ: IEEE.
2. Ning, Lu, Hua Lin, Lu Jie, and Guangquan Zhang. 2012. A customer churn prediction model in telecom industry using boosting. *IEEE Transactions on Industrial Informatics* 10 (2): 1659–1665.
3. Bi, Wenjie, Meili Cai, Mengqi Liu, and Guo Li. 2016. A big data clustering algorithm for mitigating the risk of customer churn. *IEEE Transactions on Industrial Informatics* 12 (3): 1270–1281.
4. Amin, Adnan, Saeed Shehzad, Changez Khan, Imtiaz Ali, and Sajid Anwar. 2015. Churn prediction in telecommunication industry using rough set approach. In *New trends in computational collective intelligence*, 83–95. Berlin: Springer.

5. Abbas, Syed Manzar, Khubaib Amjad Alam, and Kwang-Man Ko. 2020. A three-way classification with game-theoretic n-soft sets for handling missing ratings in context-aware recommender systems. In *2020 IEEE International Conference on Fuzzy Systems (FUZZ-IEEE)*, 1–8. Glasgow: IEEE.
6. Azam, Nouman, and JingTao Yao. 2014. Game-theoretic rough sets for recommender systems. *Knowledge-Based Systems* 72: 96–107.
7. Abbas, Manzar, Muhammad Usman Riaz, Asad Rauf, Muhammad Taimoor Khan, and Shehzad Khalid. 2017. Context-aware youtube recommender system. In *2017 international conference on information and communication technologies (ICICT)*, 161–164. Karachi: IEEE.
8. Abbas, Syed Manzar, Muhammad Usman Riaz, Muhammad Taimoor Khan, and Shehzad Khalid. 2017. Improved context-aware youtube recommender system with user feedback analysis. *Bahria University Journal of Information & Communication Technology* 10(2).
9. Abbas, Syed Manzar and Khubaib Amjad Alam. 2019. Exploiting relevant context with soft-rough sets in context-aware video recommender systems. In *2019 IEEE international conference on fuzzy systems (FUZZ-IEEE)*, 1–6. New Orleans, LA: IEEE.
10. Abbas, Syed Manzar, Khubaib Amjad Alam, and Shahaboddin Shamshirband. 2019. A soft-rough set based approach for handling contextual sparsity in context-aware video recommender systems. *Mathematics*, 7(8): 740.
11. Zhou, Bing, Yiyu Yao, and Jigang Luo. 2010. A three-way decision approach to email spam filtering. In *Canadian conference on artificial intelligence*, 28–39. Berlin: Springer.

QAExtractor: A Quality Attributes Extraction Framework in Agile-Based Software Development

Mohsin Ahmed, Saif Ur Rehman Khan, and Khubaib Amjad Alam

Abstract Many software projects fail due to an inadequate understanding of their quality at the initial stages of the development. The quality of the software heavily depends on non-functional requirements. But requirements elicitation team pay less attention towards non-functional requirements, during the requirements elicitation stage of software development life cycle. In agile-based software development, the requirements elicitation team only focuses on the user stories (i.e. functional requirements in an agile-based software development context). The non-functional requirements are neglected, and team members are left only with the user stories. This makes it very difficult for them to make decisions about quality while holding user stories in their hands. Furthermore, it is essential to affirm the quality of the software as early as possible. This is mainly because the quality of software can negatively affect the different artefacts of the software at later development stages, i.e. designing, developing, etc. The early stage conformance of software quality is more important in the Agile-based Software Development context, where the requirements are more volatile than any other development environments. Hence, there is a need to develop such an automated framework which can extract quality attributes from user stories automatically. In this work, we propose a Quality Attributes Extraction Framework, named as QAExtractor, for the extraction of key quality attributes from functional requirements (i.e. user stories in Agile-based Software Development context). The core of this framework is based on Natural Language Processing. The proposed technique grounds on the regular expressions, which generalise the user stories for a specific quality factor and attribute. The quality factor defines the context of the user story, i.e. security, while quality attribute states the quality aspect, i.e.

M. Ahmed (✉) · S. U. R. Khan
Department of Computer Science, COMSATS University Islamabad, Islamabad 45550, Pakistan
e-mail: its.mohsin.ahmed@gmail.com

S. U. R. Khan
e-mail: saif_rehman@comsats.edu.pk

K. A. Alam
Department of Computer Science, National University of Computer and Emerging Sciences (FAST-NUCES), Islamabad 44000, Pakistan
e-mail: Khubaib.amjad@nu.edu.pk

S. Anwar and A. Rauf (eds.), *Proceedings of the First International Workshop on Intelligent Software Automation*, Advances in Intelligent Systems and Computing 1347, https://doi.org/10.1007/978-981-16-1045-5_2

integrity. The effectiveness of the proposed technique is validated using a case study. The experimentation results revealed that the proposed framework outperforms the existing ones in terms of accuracy, precision, recall and F measure.

Keywords Software quality attributes · Automatic extraction · Agile software development · Software quality assurance

1 Introduction

Mostly the software organisations prefer functional requirements over Non-Functional Requirements (NFRs). While doing so, they pay less attention to the NFRs and neglect quality attributes, which cause failure or less quality to the project [1, 2]. These NFRs play an important role while making design decisions. The quality of the software is not a single-phase activity. It starts from the initialisation of the software project and goes throughout the stages of development. The software quality is concerned with the functional behaviour of software because it shows how much the customer is satisfied with the provided functionality. That is why the quality attributes of software are related to the functional requirements [3, 4]. The functional requirements specify the functional behaviour of software individually, and quality attributes are the properties that affect the system as an individual or whole. The quality attributes also depend upon the domain or context of the software and specific development phase [5]. The required quality level of the software is an obstacle in its development [6].

In Agile-based Software Development (ASD), the requirements are written or recorded as user stories [5, 6]. The user stories are more uncertain than the functional requirements in other development environments, and it is more difficult to specify the quality attributes from them. There are some specific issues with the selection of quality attributes, based on user stories. First, the Requirements Elicitation Team (RET) mostly ignore the quality aspect of the software while eliciting the requirements in an agile context [7]. They neglect NFRs and focus on the elicitation of user stories. Second, the quality attributes vary for the different stakeholders. Even, they change for a single group of stakeholders, e.g. one user says that the requirement "A" is most important while another user says that it is least important [8]. Different stakeholders have their own concerns about the quality of the software system. There is a strong link between the types of stakeholders and software quality [9]. So, the software requirements are classified into the levels to specify their priority. Some Quality attributes are overt or implicit such as functionality and usability. On the other hand, some attributes are optional or variant, such as learnability, efficiency, respectively.

Moreover, some of the quality attributes conflict with each other that can cause issues in design decisions. These issues can lead to failure or an unsuccessful software system [4]. Once the quality attributes are correctly identified, they will help to make appropriate design decisions and reduce the rework which is mostly caused due to the false identification of quality attributes. Few current state-of-the-art techniques focus on the extraction of quality attributes. The ACRUM methodology [10], incremental prototyping [11] and enhanced approach [12] are manual ones. They require expert analysis to specify the quality of software, which is very time taking and costly as the experts are not freely available. On the other hand, NORMAP methodology [13] and SpaCy classifier [14] are semi and fully automated, respectively. NORMAP methodology involves in the later design phases to map quality attributes, and SpaCy classifier only focuses on two quality attributes (i.e. security and reliability) that are insufficient to define the quality of a software system.

Hence, there is a need for such an automated framework that can extract quality attributes at the early stage of software development, considering the agile context. The reason to deal with the quality attributes of software at an early stage is that as soon as these will be identified, the design decision phase will be more appropriate according to the demand of the software. In this work, we are proposing a Quality Attributes Extraction Framework, named as QAExtractor, in ASD. This framework is based on Natural Language Processing (NLP), which is a linguistic and cognitive science approach [15]. As the user stories are more informal than the requirements in other development contexts, the NLP is more suitable to automate the process of Software Quality Assurance (SQA) in ASD. In our proposed model, the regular expressions are the core of NLP. The extracted quality attributes will help to develop a new quality model for that specific software. Most of the quality models have their critical view on software quality [16]. Our designed model will give a dynamic view of the quality of software because it will be based on the extracted quality attributes of a software system. Our research contributions are

1. We investigated and classified current state-of-the-art quality attributes extraction techniques.
2. We proposed an automated framework for the extraction of quality attributes in ASD context.
3. We performed experimentation to validate the effectiveness of our proposed framework.

The rest of the paper is organised as follows: Sect. 2 contains a detailed review of existing state-of-the-art quality attributes extraction techniques in agile context, Sect. 3 presents the proposed framework and briefly explains its components, Sect. 4 shows the experimental results and effectiveness of the proposed framework, and Sect. 5 concludes the whole research work with some future directions.

2 Literature Review

In the literature, there are very few techniques that focused on the identification or extraction of quality attributes of the software project in the early stages of the development. We have analysed state-of-the-art techniques in agile context and classified them into Algorithmic and Non-Algorithmic techniques.

2.1 Algorithmic Techniques

Farid [13] presented a methodology for the modelling of NFRs, named NORMAP, in which the author specifically tailored the NFRs modelling framework for Agile practices. They used Chung's NFRs framework [17] for linking NFRs with the functional requirements, in the form of coded colours, where they represented functional requirements as Agile Use Cases (AUC) and NFRs as Agile Loose Cases (ALC). Hence, they involved in the designing phase of the SDLC. They achieved the classification success of 87.15% and their framework can be used as a manual or semi-automated, i.e. NFRs Modelling for Agile Manual (NORMANUAL) and NFRs Modelling with Agile Automatic (NORMATIC) [18]. However, this framework does not map NFRs with functional requirements at the stage of requirements elicitation and involves in the designing phase of the project.

Gilson et al. [14] presented a study in which they investigated whether either user stories contain information about quality attributes or not. They found that several quality attributes such as compatibility, security, usability, maintainability, performance and many more can be extracted from them. They utilised the Machine Learning (ML)-based classifier using the SpaCy library that is an extension of Google's universal part-of-speech tag-set [19]. The base of SpaCy library is on Convolutional Neural Network (CNN) [20], and it is pretty much reliable in terms of performance and learnability [21]. First, they implemented one global model trained for all Quality attributes. Second, they applied specialised models trained for individual quality attributes. The experimental results showed that the global model outperformed the specialised models. However, except compatibility and maintainability, these models were not tested with other quality attributes. Furthermore, the model requires incremental training to obtain better accuracy in terms of quality attributes' extraction, which is a time-consuming process.

2.2 Non-algorithmic Techniques

Jeon et al. [10] proposed a quality attribute driven agile development method named ACRUM, complying with the core activities of the SCRUM and its agility intact, to analyse the quality attributes and manage their traceability. In their proposed model, they incorporated the aspect of NFRs (i.e. quality attributes) along with the functional requirements which were missing in the SCRUM process due to its flexible nature. The functional requirements were analysed to maintain the product backlog through which the quality attributes were mapped. After implementing the sprint backlog, they verified functional requirements and quality attributes by a demonstration.

Bellomo et al. [11] focused on the quality attributes of the software by incremental prototyping where the requirements were evolving. Firstly, the feature was developed, and then early feedback was taken from the relevant stakeholder after the sprint. An architectural trade-off is designed and discussed with the stakeholder at the post-user demo, and if the stakeholder agreed on the functionality and quality of the feature, then it is approved, otherwise rework is being done. Later, that prototype-module was integrated with the actual project at release planning.

Aljallabi and Mansour [12] summarised two existing approaches for the analysis of NFRs, combined their advantages and shortcomings, and came up with an enhanced approach to NFRs analysis. They considered both perspectives of NFRs, i.e. internal and external quality attributes, to guarantee their correct identification. This approach was comprised of six phases. In phase 1, they elicited functional requirements from the users. In phase 2, they prepared a mapping sheet between functional requirement and external quality attributes where the developer used his technical knowledge to include and exclude appropriate external quality attributes according to the elicited functional requirements. In phase 3, the sheet is handed over to the customer and asked to map each functional requirement with the required quality attributes. In phase 4, internal quality attributes are decided between customer and project members. In phase 5, a mapping sheet between external and internal quality attributes is created to remove the conflict between them. Finally, this mapping sheet is discussed with the customer and finalised the attributes after the discussion.

Jawad and Bashir [22] developed an interrelationship among quality attributes and prioritised them qualitatively on the basis of power and dependency among them. Jain et al. [23] used Interpretative Structural Modelling (ISM) and presented a framework to model and measure the quality attributes in such a way that most critical attributes can be focused on. However, the identification of quality attributes was made with the subjective analysis of experts, i.e. expert's opinion.

3 Proposed Framework

To overcome the problem of manual extraction of quality attributes, we are proposing QAExtractor, a Quality Attributes Extraction Framework, to extract Quality attributes automatically from user stories. The significance of our conceptual framework is that it is an automated and early phase quality attributes extraction framework which extracts quality attributes from the user stories, at the stage of requirements elicitation of SDLC. We have selected NLP as the core component of the extractor that is based on regular expressions. The reason behind the utilisation of NLP is that the user stories are in the format of natural language. They are less formal and represent a complete story, told by the user. The NLP helps to understand the semantic. Once the semantic will be clear, the quality attribute will be extracted confidently. The proposed framework is presented in Fig. 1. It is comprised of three layers, i.e. Input Layer, Processing Layer and Output Layer.

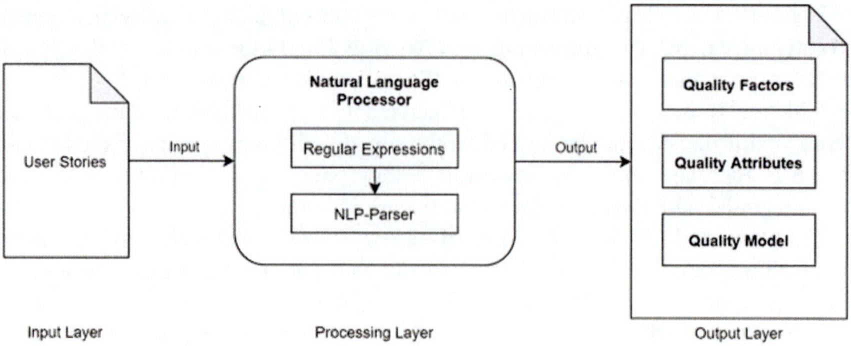

Fig. 1 Proposed framework

3.1 *Input Layer: User Stories*

User stories are the functional requirements of the software that are stated by the user. Though they are not directly stated to specify the quality of software. But they have some hidden indications that could specify the quality. These user stories are passed to the Natural Language Processor to scan them one by one and find the hidden quality attribute inside them. The quality attributes could be either explicitly stated or not. The standard format for a user story is "As a ⟨ type of user ⟩, I can ⟨ perform some task ⟩, so that I can ⟨ achieve some goal ⟩" [24]. It is not necessary for the user story to be in a specific format. However, since the technique is particularly designed for the ASD, so the validation module of the Natural Language Processor only validates the user story if it follows the specific format.

3.2 *Processing Layer: Natural Language Processor*

The Natural Language Processor is the core of the system. It is comprised of two components, Regular Expressions and NLP-Parser. It takes user stories as input and fetches the predefined regular expressions to parse them. Then, it passes the user stories one by one through these regular expressions. If the user story successfully parses through the regular expression, then it fetches the context and corresponding quality attribute of a user story and stores it. After parsing user stories, it organises all the identified contexts and quality attributes, and develops a quality model to give a better overview of extracted quality attributes.

3.2.1 Regular Expression

Regular expressions are the search patterns that are used to search the specific keywords in user stories. We cannot search these keywords directly because they are compound keywords and have more than one order. Moreover, a word could have multiple synonyms and it could be in different forms such as verb and noun. So, there is a need to design regular expressions for different quality attributes to spot them in user stories. There could be more than one regular expressions for one quality attributes as it could be expressed in different ways. On the basis of these regular expressions, the context and quality attributes are extracted from the user stories and a quality model is built.

3.2.2 NLP-Parser

NLP-Parser is responsible to fetch the regular expressions and parse the user stories through it to identify their syntactic structure. If a user story successfully parses through the regular expressions, then it fetches their contexts and quality attributes, and store them. The pseudo-code of the algorithm, for parsing user stories document and giving extracting quality attributes, is given below

Algorithm 1: User stories parsing algorithm

```
Input: User Stories Document
Result: Quality Factors, Quality Attributes
extractedQualityAttributes = null;
userStories = scanUserStories(userStoriesDocument);
for userStories 1 to m do
    userStory = userStories[m];
    for regularExpressions 1 to n do
        regularExpression = regularExpressions[n];
        if parse(userStory, regularExpression) === true then
            extract quality attribute, i.e. the index name of the regular expression
            and store it in an array;
            qualityAttribute = indexOf(regularExpression);
            push qualityAttribute into extractedQualityAttributes;
        else
            continue;
        end
    end
end
Return: extractedQualityAttributes;
```

3.3 *Output Layer*

Output layer gives the final results after processing the user stories. It includes quality factors and quality attributes that have been identified during the parsing of user stories. Each user story is parsed individually and appropriate quality factor and the quality attribute is attached with it if found during parsing. It is also possible that a user story has more than one quality factors or quality attributes. Similarly, it is also possible that one quality factor or quality attributed is found in more than one user stories. In the end, all quality factors and quality attributes are combined to generate a specific quality model.

4 Quality Attributes Extraction Method

A flow chart of quality attributes' extraction method is depicted in Fig. 2. It initiates with the input of user stories' document. As the document is provided, it parses the whole document, gets the user stories and leaves the rest of the data. The regular expressions are designed to verify either the user stories are in the standard format or not. Once, the user stories will be according to their standard format, they will be collected from the document.

For example, a user story with a standard format is shown in Fig. 3a. The regular expression for it's validation is also shown in Fig. 3b. The output after parsing user story through the regular expression is also shown in Fig. 3c. The standard structure for a user story is highlighted to illustrate the parsing results. We are slightly mod-

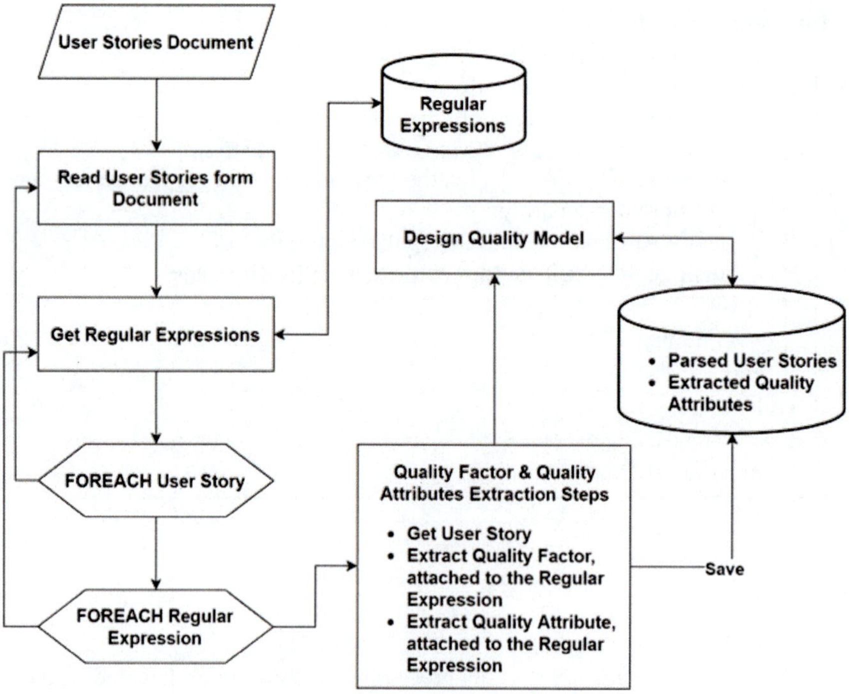

Fig. 2 Quality attributes extraction method—flow chart

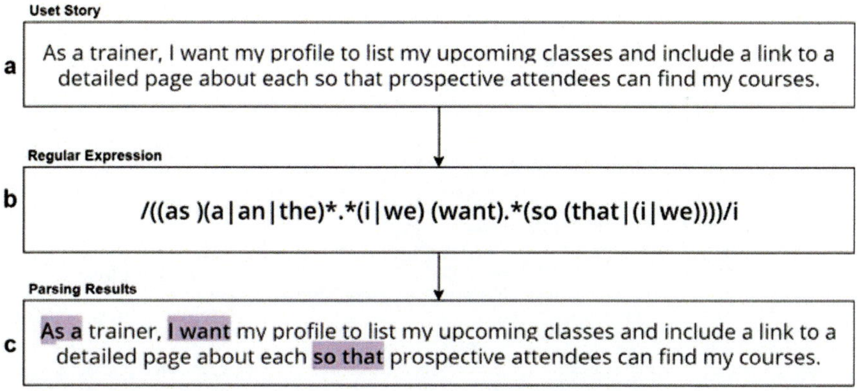

Fig. 3 Validation of user story

Fig. 4 Regular expressions stored in a multi-dimensional associative array

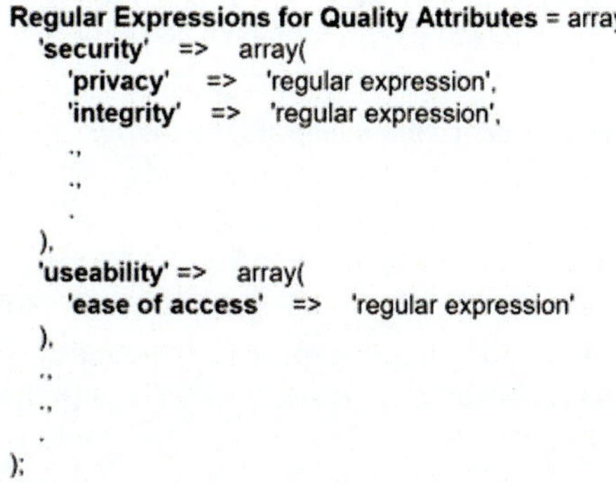

ifying the regular expression to accommodate the different articles (i.e. a, an, the) and pronouns (i.e. I, we).

After getting all user stories from the document, the extraction method gets regular expressions to extract quality attributes from the user stories. It reads the collected user stories one by one, parses each user story through all regular expressions and checks either the user story is successfully parsed through any regular expression or not. The regular expressions are stored in the form of a multi-dimensional associative array and the names of the index of this array are the quality factors and quality attributes, as shown in Fig. 4. Once, the user story is successfully parsed through the regular expression, it extracts the attached quality factor and quality attribute to the regular expression, and stores them. When all user stories are successfully checked against all regular expressions, the extracted quality factors and quality attributes are gathered and organised to design a quality model.

A working illustration of QAExtractor is shown in Fig. 5. As a sample, a user story (i.e. from the dataset, used for the experimentation) is shown in Fig. 5a. A multi-dimensional associative array of regular expressions is shown in Fig. 5b. For the illustration, only one regular expression is shown. Finally, the parsing results of

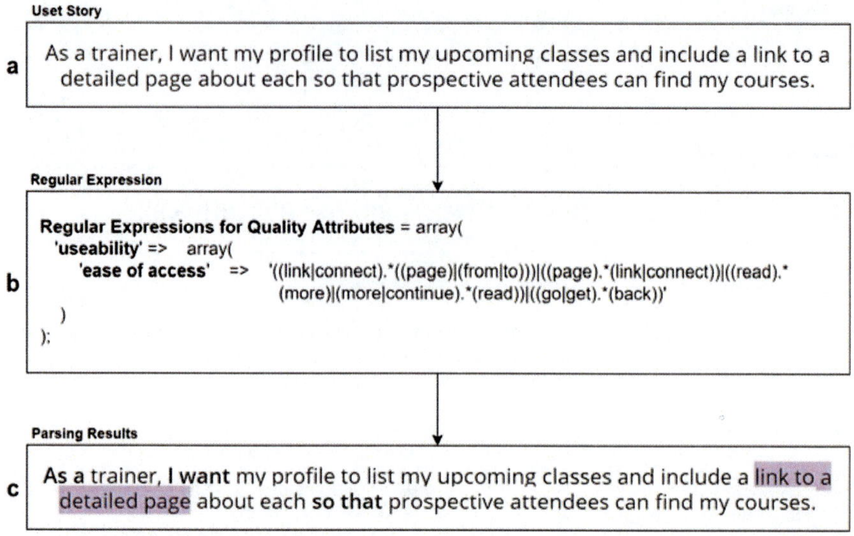

Fig. 5 Working illustration of QAExtractor

a user story are shown in Fig. 5c. The phrase that successfully parsed through the given regular expression is highlighted. Since the user story is successfully parsed through the regular expression, the indexes of the associative array are collected as quality attributes, i.e. useability and ease of access.

5 Experimental Results

The proposed framework is implemented and validated on an online server, using PHP 5.0 as a server scripting language. We used an open-source online dataset of user stories [25]. We provided the user stories' document as an input, the framework processed user stories one by one and extracted quality factors and quality attributes automatically. In the end, we generated a custom quality model on the basis of extracted quality factors and quality attributes. Our proposed system successfully extracted 6 quality factors and 15 quality attributes. These quality factors and quality attributes are given in Table 1.

While testing the proposed system, we found that one user story can have more than one quality factors or quality attributes. Similarly, one quality factor or quality attribute can occur in more than one user stories. After parsing 148 user stories, we are showing the frequency of extracted quality attributes in Table 2.

We hired a quality assurance expert from the industry to label the user stories to benchmark the effectiveness of our proposed framework. On the basis of these labelled user stories, we formed a confusion matrix by calculating true positives, true negatives, false positives and false negatives. The confusion matrix is given in Table 3.

Table 1 Successfully extracted quality factors and quality attributes

Quality factor	Quality attribute
Security	Security Integrity Authenticity
Useability	Ease of access Operability Synthesis
Reliability	Correctness Completeness Consistency Stability
Maintainability	Flexibility Modularity Modifiability
Efficiency	Accessibility Recovery
Understandability	Consistency Modularity

Table 2 Frequency of extracted quality attributes

Quality attribute	Frequency
Security	5
Integrity	21
Authenticity	1
Ease of Access	172
Operability	7
Synthesis	6
Correctness	13
Completeness	24
Consistency	72
Stability	43
Flexibility	17
Modularity	1
Modifiability	3
Accessibility	9
Recovery	2

Table 3 The confusion matrix formed by experimental results

n = 407	Actually positive	Actually negative
Predicted positive	True positives (297)	False positives (76)
Predicted negative	False negatives (26)	True negatives (8)

Table 4 Performance metrics of the results

Technique	Accuracy	Precision	Recall	F1 score
SpaCy classifier [14]	–	0.74	0.42	0.53
QAExtractor	0.75	0.79	0.92	0.85

Once we got all these values, we quantified the effectiveness of the QAExtractor Framework by calculating accuracy, precision, recall and f1 score. The results of these performance metrics are given in Table 4. We compared our results with an existing technique [14] and found that our technique comparatively outperformed in terms of all metrics that are used for the performance evaluation.

6 Threats to Validity

We tested the proposed framework only on a small and open-source dataset. These results might differ when the framework is applied to a medium or large-size dataset. Though the selected dataset covered different aspects of real software systems, we could not generalise the results because there are numbers of types of software systems. There is a need to be very much careful while applying this framework on complex and critical systems, such as safety-critical systems where a minor mistake could cause a huge loss or damage. Moreover, the dataset was labelled by the expert from the industry. We hired an expert from the industry and did not label the dataset by ourselves. It could also lead towards the biasness because the quality is a personal aspect which represents someone's intent and it could vary person by person. Furthermore, the proposed framework can extract the above-mentioned 15 quality attributes, as shown in Table 1. If a software system has any other hidden quality attribute, then the proposed framework would not be able to extract it.

7 Conclusion and Future Directions

The research is very limited in extracting or identifying quality attributes from functional requirements of the software, specifically at the initial stage of software development in an agile context. We proposed an NLP-based Quality Attributes Extraction Framework for automatic extraction of key quality attributes from user stories. Our proposed framework successfully identified and extracted 15 key quality attributes. The experimental results showed considerable effectiveness in terms of accuracy, precision, recall and F1 score, i.e. 0.75, 0.79, 0.92 and 0.85 respectively. Our proposed framework automatically extracts quality attributes from user stories which significantly reduced the time complexity and minimised the need of experts. In future, we will be focusing on refining the regular expressions to enhance the effectiveness of QAExtractor. Furthermore, we will also focus on the detection of individual quality attributes to improve the whole framework.

References

1. Domah, Darshan. 2013. The NERV methodology: non-functional requirements elicitation, reasoning and validation in agile processes. PhD thesis, Nova Southeastern University.
2. Maiti, Richard R and Frank J Mitropoulos. 2017. Capturing, eliciting, and prioritizing (CEP) NFRs in agile software engineering. In *SoutheastCon 2017*, 1–7. Charlotte: IEEE.
3. Moreira, Ana, João Araújo, and Isabel Brito. 2002. Crosscutting quality attributes for requirements engineering. In *Proceedings of the 14th international conference on Software engineering and knowledge engineering*, 167–174.
4. Al Imran, Md Abdullah, Sai Peck Lee, and MA Manazir Ahsan. 2017. Measuring impact factors to achieve conflict-free set of quality attributes. In *2017 IEEE 8th control and system graduate research colloquium (ICSGRC)*, 174–178. Shah Alam: IEEE.
5. Arvanitou, Elvira Maria, Apostolos Ampatzoglou, Alexander Chatzigeorgiou, Matthias Galster, and Paris Avgeriou. 2017. A mapping study on design-time quality attributes and metrics. *Journal of Systems and Software* 127: 52–77.
6. Malik Hneif and Sai Peck Lee. 2010. Using guidelines to improve quality in software nonfunctional attributes. *IEEE Software* 28 (6): 72–77.
7. Chung, Lawrence and Julio Cesar Sampaio do Prado Leite. On non-functional requirements in software engineering. In *Conceptual modeling: Foundations and applications*, 363–379. Berlin: Springer.
8. Etxeberria, Leire, Goiuria Sagardui Mendieta, and Lorea Belategi. 2007. Modelling variation in quality attributes. *VaMoS* 7: 51–59.
9. Weilemann, Erica and Philipp Brune. 2020. The influence of personality on software quality—a systematic literature review. In *Trends and innovations in information systems and technologies*, ed. Rocha, Álvaro, Hojjat Adeli, Luís Paulo Reis, Sandra Costanzo, Irena Orovic, and Fernando Moreira, 766–777. Cham: Springer International Publishing.
10. Jeon, Sanghoon, Myungjin Han, Eunseok Lee, and Keun Lee. 2011. Quality attribute driven agile development. In *2011 Ninth international conference on software engineering research, management and applications*, 203–210. Baltimore, MD: IEEE.
11. Bellomo, Stephany, Robert L Nord, and Ipek Ozkaya. 2013. Elaboration on an integrated architecture and requirement practice: Prototyping with quality attribute focus. In *2013 2nd international workshop on the twin peaks of requirements and architecture (TwinPeaks)*, 8–13. San Francisco, CA: IEEE.
12. Aljallabi, Bahiya M and Abdelhamid Mansour. 2015. Enhancement approach for non-functional requirements analysis in agile environment. In *2015 international conference on computing, control, networking, electronics and embedded systems engineering (ICCNEEE)*, 428–433. Khartoum: IEEE.

13. Farid, Weam M. 2012. The normap methodology: Lightweight engineering of non-functional requirements for agile processes. In *2012 19th Asia-Pacific software engineering conference*, vol 1, 322–325. Hong Kong: IEEE.
14. Gilson, Fabian, Matthias Galster, and François Georis. Extracting quality attributes from user stories for early architecture decision making. In *2019 IEEE international conference on software architecture companion (ICSA-C)*, 129–136. IEEE.
15. Collobert, Ronan, Jason Weston, Léon Bottou, Michael Karlen, Koray Kavukcuoglu, and Pavel Kuksa. 2011. Natural language processing (almost) from scratch. *Journal of Machine Learning Research*, 12: 2493–2537.
16. Nistala, P., K. V. Nori, and R. Reddy. 2019. Software quality models: A systematic mapping study. In *2019 IEEE/ACM international conference on software and system processes (ICSSP)*, 125–134.
17. Chung, Lawrence, Brian A Nixon, Eric Yu, and John Mylopoulos. 2012. *Non-functional requirements in software engineering*, vol 5. Berlin: Springer Science & Business Media.
18. Farid, Weam M and Frank J Mitropoulos. 2012. Normatic: A visual tool for modeling non-functional requirements in agile processes. In *2012 proceedings of IEEE Southeastcon*, 1–8. Orlando, FL: IEEE.
19. Petrov, Slav, Dipanjan Das, and Ryan McDonald. 2011. A universal part-of-speech tagset. arXiv:1104.2086.
20. Yin, Wenpeng, Katharina Kann, Mo Yu, and Hinrich Schütze. 2017. Comparative study of CNN and RNN for natural language processing. arXiv:1702.01923.
21. Al Omran, Fouad Nasser A and Christoph Treude. 2017. Choosing an NLP library for analyzing software documentation: a systematic literature review and a series of experiments. In *2017 IEEE/ACM 14th international conference on mining software repositories (MSR)*, 187–197. Buenos Aires: IEEE.
22. Jawad, Asil Nehad Abdel and Bashir Hamdi. 2015. Hierarchical structuring of organizational performance using interpretive structural modeling. In *2015 international conference on industrial engineering and operations management (IEOM)*, 1–7. Dubai: IEEE.
23. Jain, Parita, Arun Sharma, and Laxmi Ahuja. 2016. Ism based identification of quality attributes for agile development. In *2016 5th international conference on reliability, Infocom Technologies and Optimization (Trends and Future Directions)(ICRITO)*, 615–619. Noida: IEEE.
24. Raharjana, Indra Kharisma, Daniel Siahaan, and Chastine Fatichah. 2019. User story extraction from online news for software requirements elicitation: A conceptual model. In *2019 16th international joint conference on computer science and software engineering (JCSSE)*, 342–347. Chonburi: IEEE.
25. User stories and user story examples by mike cohn. https://www.mountaingoatsoftware.com/agile/user-stories. (Accessed 23 Sep 2020).

Automated Classification of Mobile App Reviews Considering User's Quality Concerns

Bisma Rehman, Khubaib Amjad Alam, and Kwang Man Ko

Abstract App review analysis has recently emerged as an active research area in software engineering considering the immensely large user base and potential benefits of automated information extraction. To achieve user satisfaction and to survive in the app market, addressing these requirements is essential. Therefore, in this research, we formulate this problem as a Multi-label classification problem and propose a classification model. We have trained multi-label review classifier-convolutional neural network (MLRC-CNN) model on top of pre-trained word embedding (known as word2vec embedding) for review level classification of ten thousand user reviews extracted from fifty Android apps. This is the first study of its kind in mobile app review classification that classifies mobile app reviews, using CNN classifier, considering user's quality concerns related to an app. Considering the promising results and significance of this model, it can be exploited as a classification model for other user feedback analysis platforms.

Keywords ISO25010 · Natural language processing · Convolutional neural network

1 Introduction

User feedback is considered as an immense source of knowledge, contains information that express user's concerns related to the quality of mobile applications. Analyzing user reviews posted on mobile app distribution platforms such as Google

B. Rehman (✉) · K. A. Alam
Department of Computer Science, National University of Computer and Emerging Sciences (FAST-NUCES), Islamabad 44000, Pakistan
e-mail: f180818@nu.edu.pk

K. A. Alam
e-mail: khubaib.amjad@nu.edu.pk

K. M. Ko
Department of Computer Engineering, Sangji University, Wonju-si, Republic of South Korea
e-mail: kkman@sangji.ac.kr

S. Anwar and A. Rauf (eds.), *Proceedings of the First International Workshop on Intelligent Software Automation*, Advances in Intelligent Systems and Computing 1347, https://doi.org/10.1007/978-981-16-1045-5_3

play store is a challenging problem due to the variability of users' expression about quality aspects of an app in user reviews [1, 2]. To measure the quality of any software product, standard quality models are defined by 'International Organization for Standardization' (ISO). Different versions of these models are proposed at different times, the latest product quality model is ISO25010, proposed in 2011. This model describes two aspects of product quality evaluation, the first is 'Product Quality Model' and the second is 'Quality in Use Model'. Product quality incorporates features, that make the product free from defects and deficiencies, have tendency to meet customer needs and improve products to satisfy consumers. Moreover, 'Quality in use' defined as users' perspective of the product quality by considering the end results of the usage of a software product instead of the software properties, it is a collective effect of the quality characteristics of a software product [3]. These quality attributes contain non-functional requirements (NFRs) that are usually overlooked during Requirement Elicitation phase, due to the main focus on getting functional features of a system which are explicitly defined. The user reviews are natural language text, the information related to non-functional attributes is latent within that text, which can lead to the lack of effective and efficient modelling methods and vague understanding of NFRs [4]. To extract information from the natural language text of user reviews, an automated classification approach is required such as MLRC-CNN. Up to now, various approaches have been used for automatic extraction of information from user reviews. However, they differ in terms of datasets or evaluation strategies, which makes the accuracy comparison of these methods impractical.

In recent years, the deep learning models performed effectively in speech recognition [5] and computer vision [6–8] and composition is performed over these learned vectors for classification [9]. Word vectors are basically the feature extractors, in which words are projected from sparse to a low-dimensional vector space through a hidden layer with 1-of-V encoding where V is a vocabulary size. In low-dimensional vector space, the dense representation has semantically similar words that are closer to each other in terms of cosine similarity or Euclidean distance [10]. Convolutional neural network (CNN) applies convolving filters to local features using stack of different layers [11]. Initially designed for computer vision but also shown subsequently effective results for NLP and semantic parsing [7], sentence modelling [12], other traditional NLP tasks [9] and search query retrieval [13]. In this study, we train a CNN classifier with one-dimensional convolution layer on top of word vectors. These vectors are publically available, obtained from an unsupervised model of neural language and were trained on 100 billion words of Google News by Mikolov et al. [6]. At first, we train the model with other parameters while keeping word vectors static. This simple model achieves the excellent results to classify reviews into multiple labels. Then learning the vectors related to the task through fine tuning also showed excellent results. This work is inspired from the Kim-CNN for sentence classification [14].

2 Model

The architecture of the multi-label-review-classifier CNN (MLRC-CNN) model shown in Fig. 1 is variant of CNN for text classification by Kim. In a review sentence, the i-th word represents the d-dimensional word vector when $x_i \in \mathbb{R}^d$. A review sentence is padded where necessary to length r, represented as $x_{1:n} = x_1 \oplus x_2 \oplus ... \oplus x_n$, where $\oplus$ serves as the concatenation operator. Generally, in $x_{i:i+j}$ correspond to concatenating words $x_i, x_{i+1}, ..., x_{i+j}$. A convolution is a mathematical operation that contains $w \in \mathbb{R}^{h*d}$ filter, applied to the words with window size of h to generate new features. For instance, a feature c_i is produced from a window of words $x_{i:i+h-1}$ by

$$c_i = f(w \cdot x_{i:i+h-1} + b) \tag{1}$$

Here, $b \in \mathbb{R}$ is a bias term and f is a non-linear function such as the sigmoid. This filter is applied to each possible window of words in the sentence $\{x_{1:h}, x_{2:h+1}, ..., x_{n-h+1:n}\}$ to produce a feature map

$$c = [c_1, c_2, ..., c_{n-h+1}] \tag{2}$$

with $c \in \mathbb{R}^{n-h+1}$. Later, we apply the Global-maxpooling operation over the generated feature map and take the global maximum value $\hat{c} = globalmax\{c\}$ as the feature corresponding to this particular filter. For each feature map, the purpose is to capture the most important features-the one with highest value. The pooling structure automatically deals with the variable length of sentences. Each feature is extracted from the one filter, respectively, the model uses multiple filters to obtain multiple features, with variable window sizes of length 1, 2, 3, 5 and 7. These features structure the penultimate layer (as shown in Fig. 1 and passed to dense layer with sigmoid output function, the output is the chances of occurrence of labels in form of probability. The model presents a single channel architecture. In one model variation, the word

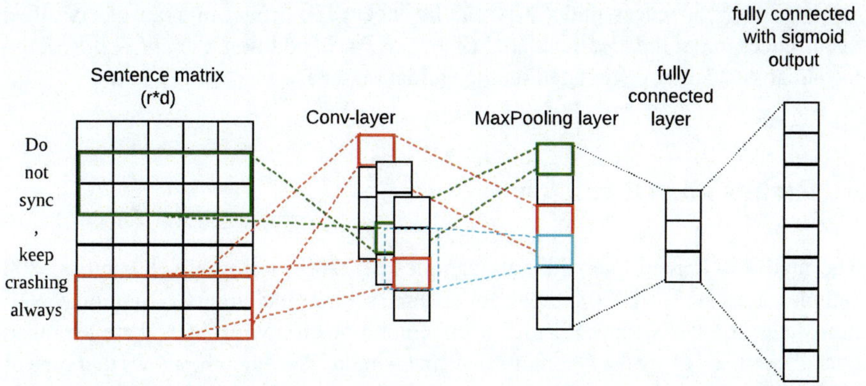

Fig. 1 MLRC-CNN architecture

vectors are fine tuned through the process of backpropagation while in the other, we kept the word vectors static throughout training.

3 Data Collection

We have collected the dataset of ten thousand reviews from Google play store considering ten applications from each of five different categories. The reviews are fetched using the *WebHarvy* tool kit. Classification involves the prediction of multiple labels, including the ISO25010 model attributes such as *Reliability, Usability, Compatibility, Security, Functional-suitability, performance, satisfaction and freedom-from-risk* as well as the *Functional Requirement or Feature request*. This is the first study that classifies mobile app reviews considering attributes of ISO25010 model into multiple labels using a deep learning classifier.

4 Experimental Setup

The overall experiment architecture is illustrated in Fig. 2. At first, the collected reviews sentences are integrated as whole review to perform review level analysis. Natural Language text needs to be cleaned and be freed from punctuation, symbols, and tokenized into token of words. A classifier is then applied to perform classification over the labelled data set of reviews.

We eliminate additional sources of randomness (Cross Validation-fold assignment, initialization of unknown word vectors, initialization of CNN parameters) to minimize the impact of above variants against other factors of randomness. Accuracy is the simple evaluation metric and computes the overall proportion of correctly classified instances. Accuracy is computed as a percentage of the numbers of correctly categorized occurrences under a particular label (TP) a total number of classified occurrences under the identical label $(TP + FP)$. In addition, the $val - accuracy$ is validation accuracy predicted on the validation dataset.

5 Results and Discussion

The model is trained over the data samples of 7200 instances, validated on 800 samples and tested on 2000 samples for all three settings including CNN-static, CNN-non-static, and CNN-rand. Results of three iterations for MLRC-CNN are averaged over all settings, using the Geometric-Mean formula. The outcome shows that type of rand-CNN with randomly initialized words output the highest training and validation

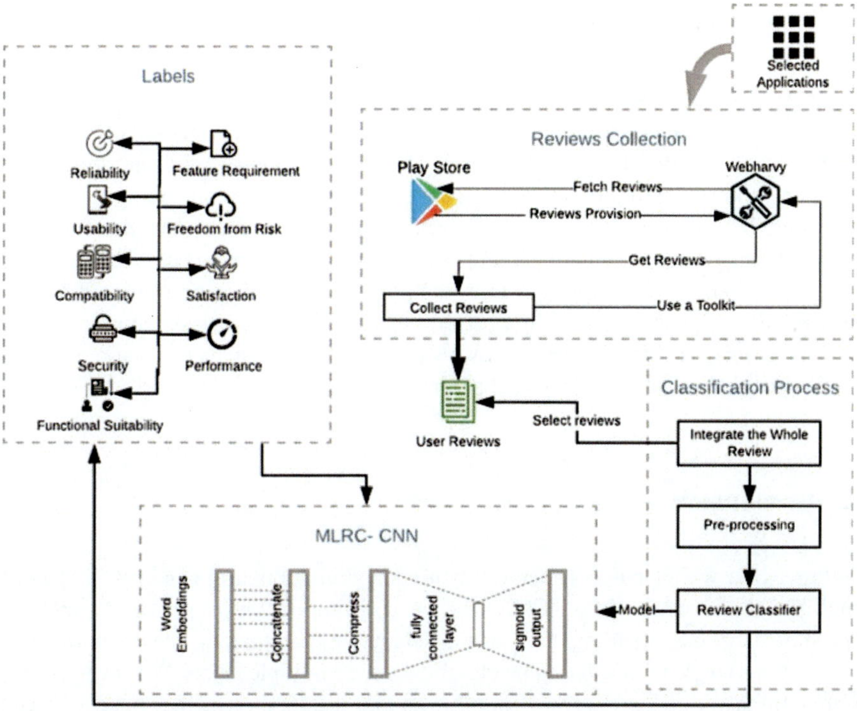

Fig. 2 Architecture of the proposed approach

Table 1 MLRC-CNN results over the dataset of ten thousands reviews

Epochs	Rand		Static		Non-static	
	Accuracy	Val_accuracy	Accuracy	Val_accuracy	Accuracy	Val_accuracy
1	0.85	0.8567	0.8544	0.8550	0.8530	0.8650
2	0.8607	0.8632	0.8552	0.8550	0.8581	0.8570
3	0.8797	0.8563	0.8555	0.8530	0.8684	0.8554
Average	0.863379581	0.858727519	0.858727519	0.858727519	0.858727519	0.858727519

accuracy. Whereas, non-static CNN has the lowest validation accuracy, while we had expected the gain in performance using pre-trained embedding. Similarly, static CNN achieved the second-highest validation accuracy and does not perform better than rand-CNN. Detailed results with all three approaches of CNN-static, CNN-non-static and CNN-rand are shown in Table 1.

6 Limitations and Future Work

MLRC-CNN effectively handles the multi-label classification of reviews. However, the number of epochs is minimal and can be increased to achieve better performance. The implementation of the MLRC-CNN classification model used the pre-trained word embedding of Google news. However, the other available embedding can also be considered. Furthermore, the context-specific embedding can be generated and used for app user reviews analysis. In future, we will consider these aspects. This research will be further extended towards more empirical case studies on other sources of user opinion, such as social media reviews and tweets. This classification model is evaluated using accuracy metrics. However, the other performance measure can also be considered to evaluate the model performance.

7 Conclusion

In this work, we have experimented with convolutional neural network to classify multi-label reviews. A simple CNN with 1D convolutional layer performs very well for multi-label classification. Our results depict that task-specific vectors performed well when a single review has to be classified under multiple labels. Word vectors are important factor in deep learning for NLP, but our results showed they have achieved accuracy less than the task-specific vectors.

References

1. Martin, William, Federica Sarro, Yue Jia, Yuanyuan Zhang, and Mark Harman. A survey of app store analysis for software engineering. 43(9): 817–847.
2. Genc-Nayebi, Necmiye and Alain Abran. A systematic literature review: Opinion mining studies from mobile app store user reviews. 125: 207–219.
3. ISO/IEC. *ISO/IEC 25010: 2011 Systems and software engineering—systems and software quality requirements and evaluation (SQuaRE)—system and software quality models*. CH: ISO Geneva.
4. Chung, Lawrence and Julio Cesar Sampaio do Prado Leite. On non-functional requirements in software engineering. In *Conceptual modeling: Foundations and applications*. Lecture notes in computer science, vol 5600, ed. Borgida, Alexander T., Vinay K. Chaudhri, Paolo Giorgini, and Eric S. Yu, 363–379. Berlin Heidelberg: Springer.
5. Graves, Alex, Abdel-Rahman Mohamed, and Geoffrey Hinton. Speech recognition with deep recurrent neural networks.
6. Mikolov, Tomas, Ilya Sutskever, Kai Chen, Greg Corrado, and Jeffrey Dean. Distributed representations of words and phrases and their compositionality.
7. Yih, Wen-tau, Kristina Toutanova, John C. Platt, and Christopher Meek. Learning discriminative projections for text similarity measures. In *CoNLL*.
8. Bengio, Yoshua, Réjean Ducharme, Pascal Vincent, and Christian Janvin. A neural probabilistic language model. 3: 1137–1155.

9. Collobert, Ronan, Jason Weston, Leon Bottou, Michael Karlen, Koray Kavukcuoglu, and Pavel Kuksa. Natural language processing (almost) from scratch.
10. Johnson, Rie and Tong Zhang. Effective use of word order for text categorization with convolutional neural networks.
11. Lecun, Y., L. Bottou, Y. Bengio, and P. Haffner. Gradient-based learning applied to document recognition. 86(11): 2278–2324.
12. Kalchbrenner, Nal, Edward Grefenstette, and Phil Blunsom. A convolutional neural network for modelling sentences.
13. Shen, Yelong, Xiaodong He, Jianfeng Gao, Li Deng, and Grégoire Mesnil. Learning semantic representations using convolutional neural networks for web search. In *Proceedings of the 23rd international conference on world wide web—WWW '14 companion*, 373–374. ACM Press.
14. Kim, Yoon. Convolutional neural networks for sentence classification. In *Proceedings of the 2014 conference on empirical methods in natural language processing (EMNLP)*, 1746–1751. Stroudsburg: Association for Computational Linguistics.

Task Scheduling in a Cloud Computing Environment Using a Whale Optimization Algorithm

Rimsha Asif, Khubaib Amjad Alam, Kwang Man Ko, and Saif Ur Rehman Khan

Abstract In cloud computing, task scheduling is the procedure of mapping the tasks onto the available resources while minimizing the execution cost and makespan. Scheduling of cloud tasks in a computing environment is a well-known NP-hard optimization problem and thus inflexible with accurate solutions. It turns out to be even more interesting in the cloud computing environment because of its heterogeneous and dynamic nature. Multiple scheduling techniques have been suggested by multiple researchers to execute the independent tasks in a heterogeneous environment of the cloud. This paper proposes the multi-objective model-based task scheduling algorithm known as the whale optimization algorithm (WOA) for optimum scheduling of independent tasks onto the available computing resources. Firstly, it uses the integer linear programming (ILP) model to calculate the fitness value by computing the makespan and execution cost function. The proposed scheduling algorithm can generate the optimum scheduler that mapped the independent tasks onto the available virtual machines (VIs) while reducing the cost and makespan. To evaluate the optimality and execution speed of the proposed scheduling algorithm, it is implemented

R. Asif · K. A. Alam (✉)
Department of Computer Science, National University of Computer and Emerging Sciences (FAST-NUCES), Islamabad 44000, Pakistan
e-mail: khubaib.amjad@nu.edu.pk

R. Asif
e-mail: f180814@nu.edu.pk

K. M. Ko
Department of Computer Engineering, Sangji University, Wonju, Republic of South Korea
e-mail: kkman@sangji.ac.kr

S. U. R. Khan
Department of Computer Science, COMSATS University Islamabad, Capital Territory, Islamabad, Pakistan
e-mail: saif_rehman@comsats.edu.pk

S. Anwar and A. Rauf (eds.), *Proceedings of the First International Workshop on Intelligent Software Automation*, Advances in Intelligent Systems and Computing 1347, https://doi.org/10.1007/978-981-16-1045-5_4

in a CloudSim environment and the simulation results show that the proposed WOA can reduce the cost and execution time in contrast to the PSO algorithm.

Keywords Task scheduling · Cloud computing · Whale optimization algorithm (WOA)

1 Introduction

Cloud computing is a promising technology that allows on-demand and convenient accessibility to a distributed set of available resources (e.g., servers, storage, network, and applications) which might be immediately allocated and de-allocated with minimum providers interaction and management services [8]. According to the NIST definition of cloud computing, it is composed of three service models: Infrastructure services (IaaS), Platform services (PaaS), and Software services (SaaS). Infrastructure services are the smallest level of abstraction supported via the CSP that gives you a complete variety of infrastructure resources, for example, storage, servers, and network hardware. [15]. Virtualization is one of the key characteristics in data centers of cloud providers that enable the sharing of computing resources, allowing multiple applications to run on dissimilar virtual machines (VMs) in a single server. By means of virtualization, service providers can validate the quality of services (QoS) delivered to dissimilar customers while achieving maximum utilization of resources and reduced communication overhead [3, 5].

In the past few years, the issue of task scheduling in a cloud computing environment has gained the huge attention of researchers. It is considered as an important issue in a computing environment by taking into consideration multiple factors such as completion time, cost of executing user's tasks, resource utilization, and power consumption [9]. Optimally allocating the independent task to the available resources is an active research area [14] and this problem is formulated as NP-complete optimization problem. Therefore, constructing a scheduler that can optimally allocate the tasks within the reasonable execution and performance bounds, is challenging in a distributed cloud environment where discovery time largely depends upon the problem size. The purpose of task scheduling [2] is to construct a scheduler that will effectively map tasks onto the computational resources.

Multiple scheduling algorithms have been suggested to resolve the issue of task scheduling in a distributed environment, for example, heuristic, meta-heuristic, hyper-heuristic, and hybrid algorithms [1]. The heuristic algorithms generate an approximate solution in a given time but don't guarantee their optimality. Conversely, the meta-heuristic algorithms are nature-inspired algorithms that can be used as a guiding strategy in designing heuristic for specific problems [13]. Hyper-heuristic is the evolving group of meta-heuristic algorithms that are combined in a way that enables the maximum utilization of the engaged meta-heuristic techniques to achieve an optimum scheduler. Whereas the hybrid algorithms are the combination of heuristic and meta-heuristic algorithms such as heterogeneous earliest finishing time (HEFT) and particle swarm optimization (PSO). It gained huge attention from researchers for their ability to discover/or find near-optimal in a sufficient amount of time.

PSO achieves enhance solutions and converges faster than ACO and GA because of its exploration ability for discovering the optimal solutions. In this article, a task scheduling technique that is constructed upon the state-of-the-art meta-heuristic whale optimization algorithm (WOA) is proposed [7] to attain optimum convergence and effective utilization of resources as compared to the PSO algorithm. The WOA algorithm does not trap in local optima solutions and provides a lesser number of parameters as compared to the PSO algorithm. Two conflicting scheduling objectives of the suggested algorithm are execution cost and makespan of executing these independent tasks onto available computational resources. The simulation results demonstrate that the suggested algorithm decreases the makespan and cost as compared to the PSO algorithm and achieves effective utilization of heterogeneous resources (VMs). The rest of the article is organized as Sect. 2 describes the related work. Section 3 formulates the system modeling. Section 4 explains the proposed scheduling algorithm (WOA). Section 5 illustrates the experimental setup on synthetic data sets. Section 6 demonstrates the conclusion and future work.

2 Related Work

Task scheduling is an NP-complete optimization and cannot be solved in polynomial time. Multiple heuristics have been proposed, for example, min-min, min-max, and minimum completion time (MCT) are employed as the best contenders for scheduling policies. As compared to this, numerous task scheduling techniques such as FCFS, SJF, and LJNF are proposed for optimization of cost and utilization of resources [1]. Furthermore, for static scheduling opportunistic balancing of load, max-min, MCT, and enhanced load balancing min-min (LBMM) are used in cloud computing [6]. A thorough explanation of multiple task scheduling techniques is demonstrated. Another heuristic is presented in [12] which combines the Max-Min and Min-Min techniques to reduce the total completion time deprived of distributing the service quality. Though the existing techniques got trapped in local optima due to the complex nature of the problem which is a multi-model issue, meta-heuristic algorithms might be employed to prevent this restriction.

The aim to employ meta-heuristic algorithms is due to their flexibility and simplicity. The majority of the meta-heuristic techniques are easy to develop, less complex, and simple. The meta-heuristic techniques might be further categorized into nature-inspired, bio-inspired, swarm intelligence, and evolutionary algorithms [11]. The genetic algorithm that is based on an evolutionary algorithm is proposed for multiple types of problems related to scheduling in cloud computing. For static scheduling of independent tasks in cloud computing, an improved GA [4] is suggested which combines the GA with the HEFT heuristic and is named as N-GA. Another author [10] enhances the execution of the GA by changing the operators which ensure the diversity and reliability of the sample space. Scheduling techniques which are constructed upon SI techniques such as PSO, ant colony optimization (ACO), and artificial bee colony (ABC), are appropriate. To further enhance the task scheduling for heterogeneous platforms, ACO-based techniques are executed for real-time scenarios. Although still there is a need to develop more effective algorithms for scheduling independent tasks in a cloud computing environment.

We presented the whale optimization algorithm for scheduling independent tasks onto the computational resources.

3 Problem Formulation

3.1 System Model

The model of the independent task scheduling in a heterogeneous computing environment is demonstrated in Fig. 1. First of all, the user submits the task (or computation to be performed) on the Internet. After that, the Application request handler addresses all the requests and transmits them to the controller node if it's legal other-wise the request is rejected. The controller node manages the request from another component of the system. All the request of legal jobs from the controller node is forward to the task queue which is further examined in accordance with the request type, for example, interactive scientific workflow application, and batch processing. After recognizing the application type from the task queue, after that, it transmits to the task analyzer module to understand the details of every application (task), for example, the coming request is based on priority or not. Mapped metrics are responsible for mapping the independent tasks onto the computing resources, while the main purpose of the scheduler is to discover the optimum allocation of tasks onto the computing resources by means of the proposed scheduling algorithms along with the defined constraints and improve the QoS. Resource monitoring node monitors the VM characteristics (VM Id, CPU, and memory) at constant intervals. We suppose that the cloud is made up of multiple physical machines (PM) and every PM consist of multiple VIs. which might be denoted as in Eq. 1:

$$CL = \{PM_1, PM_2, PM_3, \ldots\ldots, PM_{100}\} \tag{1}$$

where CL denotes the cloud and PM_1 shows the physical resources present in the cloud. The subsequent equations can present the PMs.

$$PM = \{VI_1, VI_2, \ldots VI_K, \ldots VI_{10}\} \tag{2}$$

Where $\{VI_1, VI_2, \ldots VI_K, \ldots VI_{10}\}$ represents the virtual machines accessible in the PM_1. Every VI contains the central processing unit (CPU) and the bandwidth rate (BR). However, the ability of the CPU is measured in MIPS, that is ranged

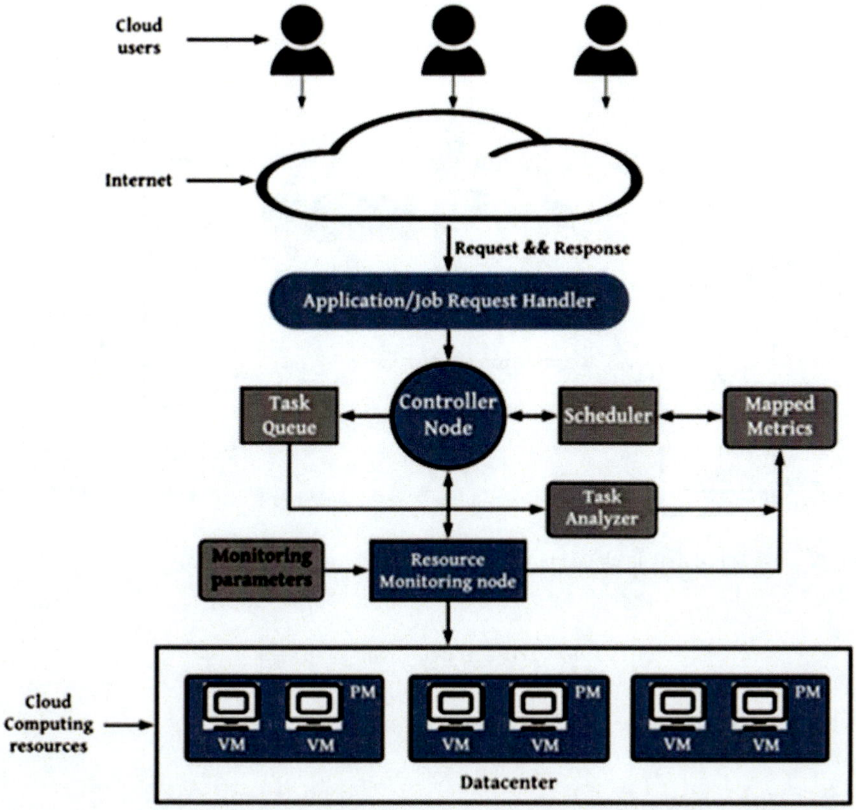

Fig. 1 System architecture of task scheduling in cloud computing environment

from 1000 to 3000 MIPS. Namely, the CPU must be able to achieve Million of instructions in one second. The total amount of tasks is 100 and every task has dissimilar requirements for both CPU, BR, deadline, and budget. Wherever ta_1 are the initial task separately without any dependency on another task, ta_j represents the jth task and ta_{100} signifies the 100th task. The main procedure of the task scheduler is explained in Fig. 2. The controller delivers the tasks to the scheduler and every task delivered by the customer includes the subsequent parameters, for example CPU, cost, BR, and QoS requirements. According to the parameters of the task scheduler, it schedules and prioritizes the tasks. As shown in Fig. 2 the CPU and BR requirements of the scheduled tasks are examined by the task scheduler and are delivered to the processing capacity of the VIs. CU_i signifies the CPU requirement of the task ta_i which signifies their total execution time. BR_i defined the bandwidth rate needed to transfer the data between different VIs.

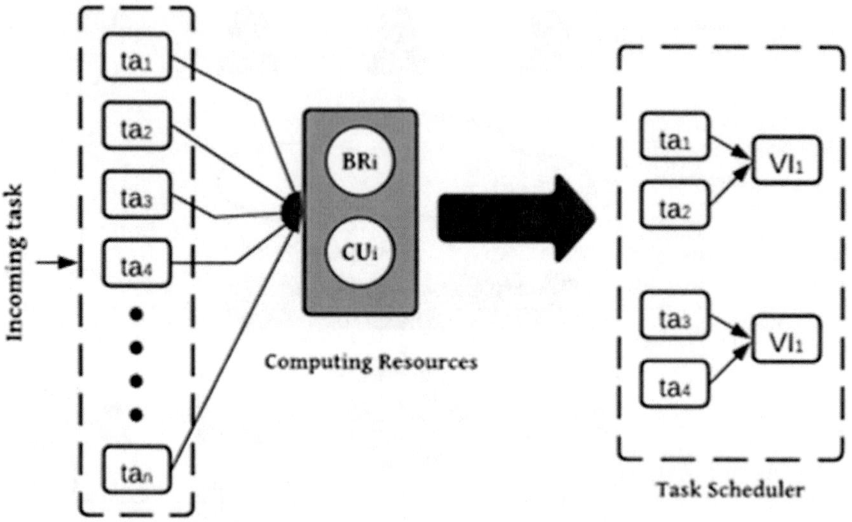

Fig. 2 Task Scheduling in cloud computing

$$Ta = \{ta_1, ta_2, \ldots, ta_j \ldots ta_{100}\} \tag{3}$$

3.2 Makespan Model

Dissimilar resources can execute diverse tasks. Processing time $PT_{i,k}$ of task ta_i on k resources is computed by the equation:

$$Processing\ time_{i,k} = \frac{Task\ Size(TS)}{Processing\ Capacity\ of\ VI_k} \tag{4}$$

There is no dependency between the tasks (tasks are independent) so the communication overhead is not considered.

$$Makespan = Max\{Processing\ time_{i,k}\} \tag{5}$$

$$F1^{min} = Makespan \tag{6}$$

The makespan of the independent tasks is described as the maximum time of executing these tasks onto the available machines and is formulated by using the equation.

3.3 Execution Cost Model

$$Ta_{EC} = \sum_{i=1}^{M} (Processingtime_{i,k}) * C_{VI} \tag{7}$$

$$F2^{min} = Ta_{EC} \tag{8}$$

The cost of a ta_j is described by means of cost to hire resources while executing the task on virtual machine VI_i. While cloud providers utilize pay per and are priced conferring to the time period consumed by the customers. C_{VI} is the charging unit of using public resources p for every apiece of time. Furthermore, the resources of service providers are expected to remain to exist in a similar region. Thus the communication and data storage cost are expected to be negligible and the bandwidth rate among these resources is expected to be approximately identical. Simply the time to communicate information between two determined tasks is considered during the experimental evaluation.

4 Proposed Task Scheduling Algorithm Based on Whale Optimization Algorithm

This section briefly explains the employed meta-heuristic algorithm for scheduling tasks in a cloud computing environment. The proposed technique is constructed upon the ILP model and the WOA technique. The ILP model first computes the objective function that is the combination of two major conflicting objectives namely makespan and execution cost. According to the calculated fitness, it assigns the VIs toward the tasks. After that, the WOA optimizer generates N random solutions. Firstly, it supposes that the current best agent is the optimum solution and executes the procedure of searching constructed upon the current optimum solution. This process is executed until the best optimum solution is found. The basic aim is to schedule the tasks onto the available computing resources however minimizing the fitness function.

4.1 Objective Function

The fitness of solutions is calculated to find the solutions with the minimum makespan and execution cost by using Eq. 9.

$$OF = \alpha * F1^{min} + (1 - \alpha) * F2^{min} \tag{9}$$

α is the cost-time balance factor in the range of [0, 1]. F1 is the Makespan function to be minimized. F2 is the total execution cost to be minimized. Firstly, the WOA algorithm calculates the fitness value by using Eq. 9.

4.2 Whale Optimization Algorithm

A state-of-the-art meta-heuristic technique which is recognized as whale optimization is developed by mirjalili and Lewis [7]. The whale optimizer is inspired by the natural behavior of the humpback whale. These whales are one of the most interesting types of species of whales. The more attractive feature is their particular hunting technique. This foraging behavior is known as the bubble-net foraging technique. They want to chase krill and smaller fishes that are near to the surface by setting up the special bubbles in a circular motion or a nine-shaped pathway. It is executed in three main steps:

- Encircling the prey
- The bubble-net attacking technique (exploitation phase)
- Search for prey (exploration phase)

For optimizing the assignment of workflow tasks onto the VM, the technique begins with a collection of random initial solutions. In the beginning it supposes that the present best applicant is the better solution and this process is iterated multiple times till the best candidate solution is found.

Initialization: In the initialization stage, the search agent population is initiated and the solution with the best fitness function is selected randomly from the population. The population of the search agent is specified as $S_j(j = 1, 2, 3, 4, \ldots\ldots\ldots i)$

Fitness computation: The fitness is computed by using Eq. 9.

Encircling the prey: The special whale in which hunting behavior is replicated is known as Humpback whales. They might determine the position of the best agent and surround them. Hence the location of the optimum solution in a search space is not known prior. It accepts that the present best fitness function applicant solution is the prey that is the target or is near to the optimal solution. Other search agents will enhance their location against the best solution. This might be described as follows:

$$\mathbf{Q} = \mid \mathbf{E}.\mathbf{S}^*(t) - \mathbf{S}(t) \mid \tag{10}$$

Wherever t denotes the present iteration, **S** signifies the vector of position and **S*** shows the vector of the best agent. **E** signifies the vector of the coefficient. The location of the present best search agent is represented in Eq. 10. Moreover, the new location of the search agent is computed by using Eq. 11.

$$\mathbf{S}(t+1) = \mathbf{S}^*(t) - \mathbf{N} \bullet \mathbf{Q} \tag{11}$$

Where **N** also denotes the vector of coefficient. Absolute value is stated as || and • denotes the element through element multiplication. The defined coefficient vectors are computed by means of Eqs. 12 and 13.

$$\mathbf{N} = 2\mathbf{n} \bullet \mathbf{m} - \mathbf{n} \tag{12}$$

$$\mathbf{E} = 2 \bullet \mathbf{m} \tag{13}$$

The **n** value is reduced from 2 to 0 and **m** represents the vector of random in [0, 1]. The **N** and **E** values are altered to go around the positions of the best search agent.

Bubble-net attacking technique: To model this manner of humpback whales mathematically, two techniques are developed as listed below, (1) Shrinking mechanism and (2) Spiral updating location.

Shrinking mechanism: In this method, the value of **N** is in the range of [−1, 1], and the current location of the agent is denoted by means of its initial location and the present best agent location.

Spiral updating location: The position of the spiral might be updated though Eq. 14:

$$\mathbf{S}(t + 1) = \mathbf{Q}' \bullet k^{cv} \bullet \cos(2\Pi v) + \mathbf{S}^*(t) \tag{14}$$

Where c denotes the constant value, and v is in the range of [−1, 1] • denotes the element through element multiplication. Q' is computed by means of Eq. 15,

$$\mathbf{Q}' = |\mathbf{S}^*(t) - \mathbf{S}(t)| \tag{15}$$

S signifies the vector of position, and $\mathbf{S}^*$ shows the best agent vector. After that the location of the search agent might be enhanced besides the method of encircling or spiral updating.

It is noted that the whale moves around the best search agent within the spiral updating path and along an encircling mechanism simultaneously. To formulate this behavior, we suppose that there is a possibility of a 50% chance to select among either the spiral updating and encircling method to upgrade the whales' position throughout the optimization.

$$\mathbf{S}(t + 1) = \begin{cases} \mathbf{S}^*(t) - \mathbf{N} \bullet \mathbf{Q} & \text{If,} \quad p < 0.5 \\ \mathbf{Q}' \bullet k^{cv} \bullet \cos(2\Pi v) + \mathbf{S}^*(t) & \text{If,} \quad p \geq 0.5 \end{cases} \tag{16}$$

Where p represents the random number in the set among [−1, 1].

Search for prey: In this, the search agent modified their position through the randomly selected agent, and it can be described as follows:

$$\mathbf{Q} = |\mathbf{E}.\mathbf{S}_{rand} - \mathbf{S}| \tag{17}$$

Algorithm 1 Whale Optimization Algorithm (WOA)

Input: Tasks and set of VMs $VI = \{VI_1, VI_2, VI_3, \ldots\ldots VI_n\}$
Output: Task scheduling solution
Initialize the Whale Population $S_i = \{i = 1, 2, 3, \ldots\ldots N\}$ with N randomly generated solutions wherever every solution satisfies the constraints.
Compute the Fitness of every solution or search agent.
$S^* =$ Best search agent.
while $t < maxIt$ **do**
  for each solution **do**
   Update $n, N, E, v,$ and p
   if $p < 0.5$ **then** ▷ **(Probability of 50 % to choose the encircling phase)**
    if $|N| < 1$ **then** ▷ **(Encircling phase)**
     Update the position of current solution using Eq. 11
    else
     if $|N| \geq 1$ **then** ▷ **(Exploration phase OR Search for prey)**
      Chose a random solution X_{rand}
      Update the position of current solution using Eq. 18
     end if
    end if
   else
    if $p \geq 0.5$ **then** ▷ **(Probability of 50 % to choose the Spiral updating phase)** ▷ **(Spiral Updating position)**
     Update the position of current solution using Eq. 14
    end if
   end if
  end for
  Compute the fitness of every solution
  Update S^* if there is a better solution
  $t = t + 1$
  Update $n, N, E, v,$ and p
end while
return S^*

$$\mathbf{S}(t+1) = \mathbf{S}_{rand} - \mathbf{N.E} \tag{18}$$

$\mathbf{S}_{rand}$ is a random position vector chosen from the current population.

Termination: If the agent moves outside of the search space, after that the $\mathbf{S}^*$ is modified and it sets the $t = t + 1$ and this procedure is iterated repeatedly till the best location of the solution is found.

The pseudo-code of the proposed scheduling algorithm (WOA) for scheduling task scheduling onto the distributed environment is illustrated in Algorithm 1. The time complexity of the WOA algorithm is O (n) where n is the total number of iterations. The basic aim of the scheduling is to allocate the tasks onto the accessible resources optimally. The scheduling of tasks depends upon the WOA algorithm; initially, the search agent population is initiated. Afterward, the search agent is evaluated by means of the ILP model. After that, the best agent from the population of the search agent is initiated and the others modified their position according to the present best agent. Then the probability of the search agent is computed; if it is less than 0.5, then the

Table 1 Simulation parameters

Parameter	Value
RAM	512–2048
Bandwidth	1 MBps
MIPS	500–1500
Population size	100
No. of VMs	1–8
No. of tasks	30–100
Maximum iteration	1000
No. of simulation runs	10
VMs monitor	Xen
Type of manager	Time_shared, space_shared

location of the agent is modified by using Eq. 11. If it is greater than 0.5, then the location of the agent is modified by using Eq. 14. The given procedure is continuous until the optimum solution is found.

5 Performance Evaluation

In this section, the results of the proposed scheduling algorithm are presented in the heterogeneous environment of cloud computing.

5.1 Experimental Setup

To assess the performance of the proposed scheduling algorithm, it was executed in a CloudSim environment. That is the toolkit for simulating scenarios is related to the cloud computing environment. A single data center is created each including two or more heterogeneous VIs. The experimental results of the proposed scheduling algorithm are compared with the PSO algorithm. Simulation parameters for performing an evaluation are shown in Table 1.

5.2 Experimental Results

In this section, the results of the proposed scheduling algorithm are described in detail. The proposed algorithm is assessed and constructed upon the scheduling objectives such as execution cost and makespan. The parametric values for PSO are adjusting as $Inertiaweight = [0.9\ 0.7]$ and learning coefficient $C1 = 1.5$. While the parametric values for the proposed scheduling algorithm (MO-WOA) are defined as $Coefficientvector = [0 - 2]$.

Makespan Figure 3 demonstrates the execution time of the proposed algorithm on iterations ranging from 10 to 500 and the population size ranging from 150 to 250. The results of the simulation show that the proposed WOA algorithm performs better as compared to the PSO for the execution time with the 30 number of tasks.

Figure 3 shows the results of the makespan with 30, 50, and 100 number of tasks. The results of the simulation show that the proposed WOA performs better as compared to the PSO algorithm.

Execution Cost Figure 4 demonstrates the execution cost of the proposed scheduling algorithm and the PSO with the 30, 50, and 100 number of tasks, and the iterations ranging from 10 to 500 number of iterations with the population size of 200 search space. The results of the execution cost for PSO and WOA have no major significant difference but the proposed scheduling algorithm reduces the execution cost of the independent tasks as compared to the PSO algorithm. Figure 4 demonstrates the execution cost with the 30, 50, and 100 number of tasks along with the number of iterations ranging from 50 to 1000 number of iterations. The results of the simulation show that the proposed WOA reduces the execution cost of the independent tasks as compared to the PSO algorithm and performs much better as compared to the other heuristic algorithms. From the experimental results Figs. 3 and 4, it is shown that the proposed scheduling algorithm (WOA) can optimally schedule the independent tasks onto the available computational resources while minimizing the execution cost and makespan.

6 Conclusion

A user-oriented technology where customers are allowed to choose hundreds of thousands of VIs resources for every task is known as cloud computing. In this, scheduling is considered to be a significant factor in executing the task. This article presents the whale optimization algorithm (WOA) for scheduling independent tasks in a distributed environment of the cloud. Firstly, the ILP model computes the fitness value by comparing the execution time and cost function. The scheduling algorithm is suggested for the optimum allocation of independent tasks onto the available VIs. It suggests that the current best solution is the optimum solution. To assess the execution speed of the suggested algorithm, it is implemented in a CloudSim environment. The results of the simulation demonstrate that it reduces the makespan and execution cost as compared to the PSO algorithm. Only two main functional objectives, makespan and execution cost, are considered in scheduling workflow applications. While the non-functional requirements such as reliability, availability, and security must be taken into consideration for future implications. Hybrid scheduling techniques can also be developed to improve the implementation of the heuristic and meta-heuristic techniques such as WOA and HEFT.

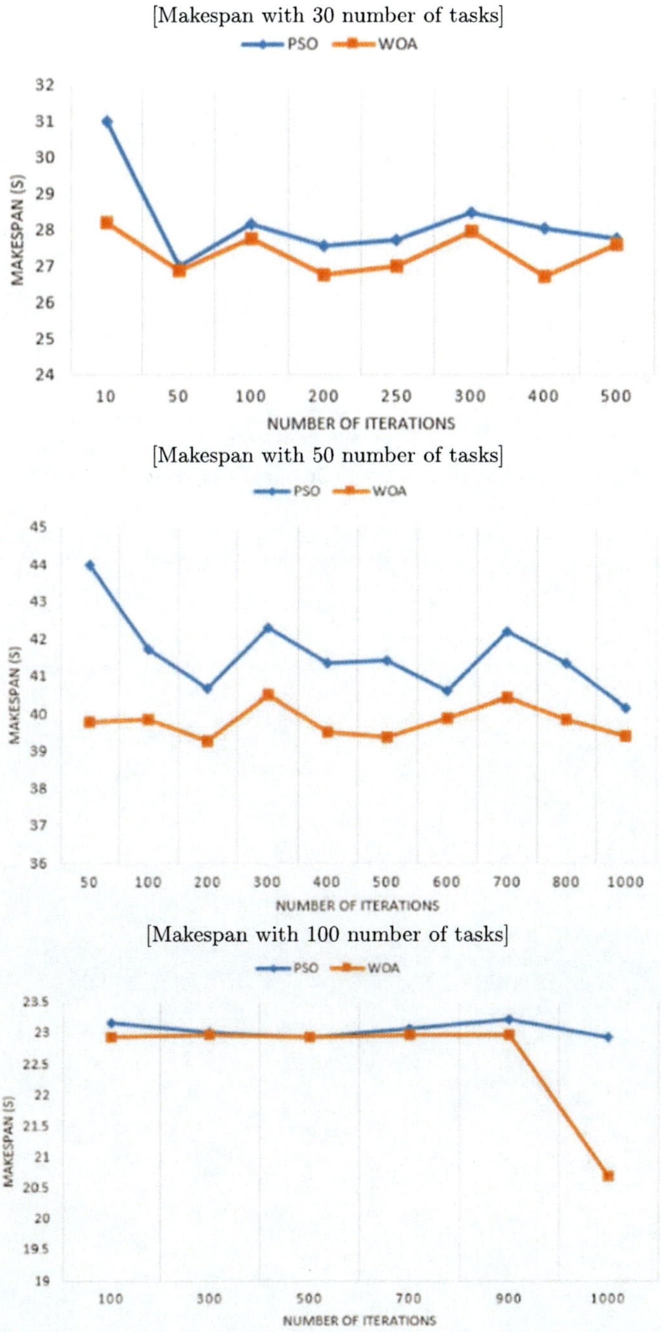

Fig. 3 Makespan

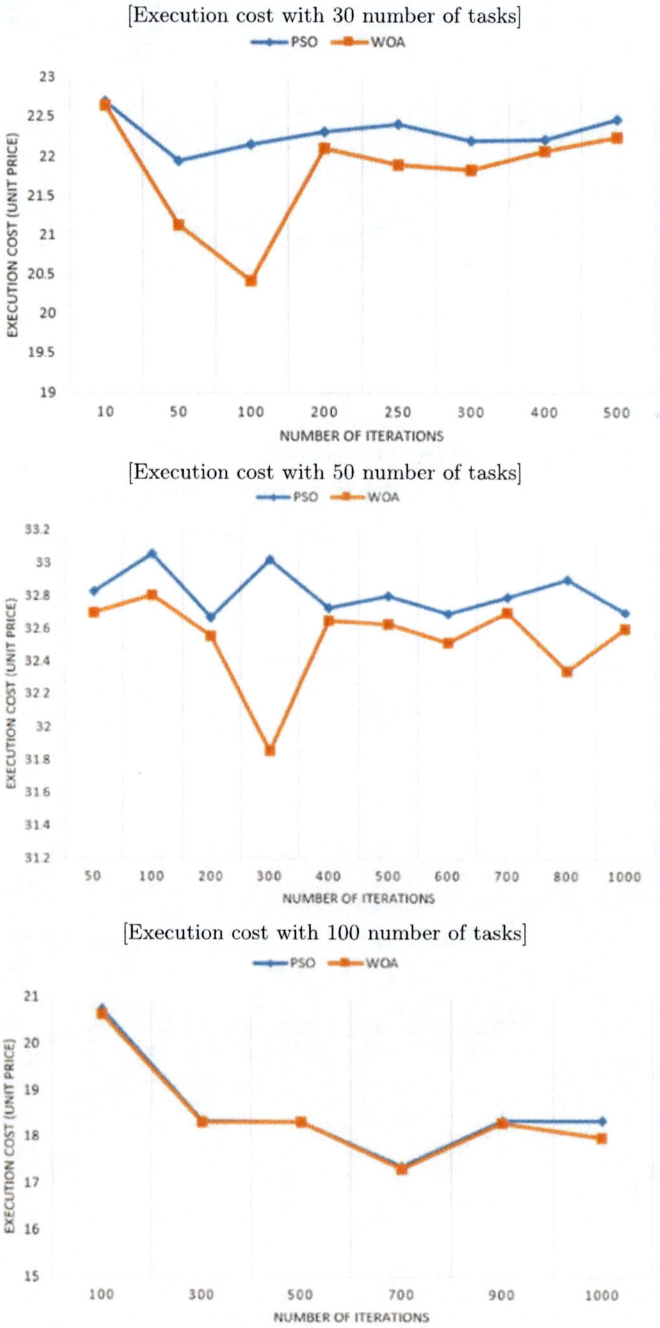

Fig. 4 Execution cost

References

1. He, Z.T., X.Q. Zhang, H.X. Zhang and Z.W. Xu. 2013. Study on new task scheduling strategy in cloud computing environment based on the simulator CloudSim. *Advanced Materials Research* 651(January): 829–834. https://doi.org/10.4028/www.scientific.net/AMR.651.829, https://www.scientific.net/AMR.651.829.
2. Juarez, F., J. Ejarque and R.M. Badia. 2018. Dynamic energy-aware scheduling for parallel task-based application in cloud computing. *Future Generation Computer Systems* 78(January): 257–271. https://doi.org/10.1016/J.FUTURE.2016.06.029, https://www.sciencedirect.com/science/article/abs/pii/S0167739X1630214X.
3. Kapil, D., E.S. Pilli and R.C. Joshi. 2013. Live virtual machine migration techniques: Survey and research challenges. In *2013 3rd IEEE International Advance Computing Conference (IACC)*, 963–969. IEEE. https://doi.org/10.1109/IAdCC.2013.6514357, http://ieeexplore.ieee.org/document/6514357/.
4. Keshanchi, B., A. Souri and N.J. Navimipour. 2017. An improved genetic algorithm for task scheduling in the cloud environments using the priority queues: Formal verification, simulation, and statistical testing. *Journal of Systems and Software* 124(February): 1–21. https://doi.org/10.1016/J.JSS.2016.07.006, https://www.sciencedirect.com/science/article/abs/pii/S0164121216301066.
5. Li, K., H. Zheng and J. Wu. 2013. Migration-based virtual machine placement in cloud systems. In *2013 IEEE 2nd International Conference on Cloud Networking (CloudNet)*, 83–90. IEEE. https://doi.org/10.1109/CloudNet.2013.6710561, http://ieeexplore.ieee.org/document/6710561/.
6. Madni, S.H.H., M.S. Abd Latiff, M. Abdullahi, S.M. Abdulhamid and M.J. Usman. 2017. Performance comparison of heuristic algorithms for task scheduling in IaaS cloud computing environment. *PloS One* 12, no. 5: e0176321. https://doi.org/10.1371/journal.pone.0176321, http://www.ncbi.nlm.nih.gov/pubmed/28467505.
7. Mirjalili, S. and A. Lewis. 2016. The Whale Optimization Algorithm. *Advances in Engineering Software* 95(May): 51–67. https://doi.org/10.1016/J.ADVENGSOFT.2016.01.008, https://www.sciencedirect.com/science/article/abs/pii/S0965997816300163.
8. Mishra, M., A. Das, P. Kulkarni and A. Sahoo. 2012. Dynamic resource management using virtual machine migrations. *IEEE Communications Magazine* 50, no. 9(September): 34–40. https://doi.org/10.1109/MCOM.2012.6295709, http://ieeexplore.ieee.org/document/6295709/.
9. Pradeep, K. and T.P. Jacob. 2018. CGSA scheduler: A multi-objective-based hybrid approach for task scheduling in cloud environment. *Information Security Journal: A Global Perspective* 27, no. 2(March): 77–91. https://doi.org/10.1080/19393555.2017.1407848, https://www.tandfonline.com/doi/full/10.1080/19393555.2017.1407848.
10. Ramezani, F., J. Lu, J. Taheri and F.K. Hussain. 2015. Evolutionary algorithm-based multi-objective task scheduling optimization model in cloud environments. *World Wide Web* 18, no. 6(November): 1737–1757. https://doi.org/10.1007/s11280-015-0335-3, http://link.springer.com/10.1007/s11280-015-0335-3.
11. Singh, P., M. Dutta and N. Aggarwal. 2017. A review of task scheduling based on meta-Heuristics approach in cloud computing. *Knowledge and Information Systems* 52, no. 1(July): 1–51. https://doi.org/10.1007/s10115-017-1044-2, http://link.springer.com/10.1007/s10115-017-1044-2.
12. Tabak, E.K., B.B. Cambazoglu and C. Aykanat. 2014. Improving the performance of IndependentTask assignment Heuristics MinMin,MaxMin and Sufferage. *IEEE Transactions on Parallel and Distributed Systems* 25, no. 5(May): 1244–1256. https://doi.org/10.1109/TPDS.2013.107, http://ieeexplore.ieee.org/document/6495454/.
13. Talbi, E.G. 2009. *Metaheuristics: From Design to Implementation*. New York: Wiley. https://www.wiley.com/en-us/Metaheuristics%3A+From+Design+to+Implementation+-p-9780470278581.

14. Tsai, C.W. and J.J.P.C. Rodrigues. 2014. Metaheuristic scheduling for cloud: A survey. *IEEE Systems Journal* 8 no. 1: 279–291 (mar 2014). 10.1109/JSYST.2013.2256731, https://ieeexplore.ieee.org/document/6516911/.
15. Yu, J., R. Buyya and K. Ramamohanarao. 2008. Workflow Scheduling Algorithms for Grid Computing. In *Metaheuristics for Scheduling in Distributed Computing Environments*, 173–214. Berlin: Springer. https://doi.org/10.1007/978-3-540-69277-5_7, http://link.springer.com/10.1007/978-3-540-69277-5_7.

Analysing GoLang Projects' Architecture Using Code Metrics and Code Smell

Maloy Kanti Sarker, Abdullah Al Jubaer, Md. Shihab Shohrawardi, Tulshi Chandra Das, and Md Saeed Siddik

Abstract GoLang or GO is a statically typed and system-level programming language created by Google. It is used for programming across large-scale network servers and big distributed systems. The measurement of the GO projects is rarely addressed in the software engineering research domain, which is needed to assess external attributes such as quality and maintainability. This paper measured two sets of widely used software metrics named Metrics For Object-Oriented Design (MOOD) and metrics proposed by Chidamber and Kemerer (CK metrics) on 5 open-sources GO projects with a range of KLOC (kilo of lines of code) [1] 7 to 601. We measured 5 MOOD and CK metrics accordingly to generate numerical results for visualizing the internal analysis. We also demonstrated the presence of two code smells named God struct and Feature envy for GO language. Finally, we have found that the majority of GO projects comprise both of these code smells with a maximum of 27 feature envy have been found which need to be refactored.

Keywords Software analytic · Code metrics · GO language · Program automation

1 Introduction

Software quality metrics concentrate on process, product and project quality elements. The goal of software quality metrics is to identify the improvement of project, planning, process and product. Various studies on software metrics distribution share the same objective of offering a way of improving the software development life cycle.

A metric is a standard unit of measure, such as meter or mile for length, or gram or ton for weight, or more generally, part of a system of parameters, or systems of measurement, or a set of ways of quantitatively and periodically measuring, assessing,

---

M. K. Sarker · A. A. Jubaer · Md. S. Shohrawardi · T. C. Das · M. S. Siddik (✉)
IIT, University of Dhaka, Dhaka, Bangladesh
e-mail: saeed.siddik@iit.du.ac.bd

S. Anwar and A. Rauf (eds.), *Proceedings of the First International Workshop on Intelligent Software Automation*, Advances in Intelligent Systems and Computing 1347, https://doi.org/10.1007/978-981-16-1045-5_5

controlling or selecting a person, process [2]. There were several studies to explain how metrics can be used to enhance the software quality [3, 4]. The calculation of a specific property of system performance or efficiency is a measure of the software metric (noun). A rule to measure a computer software object feature or attribute. Computer organizations, including specification reports, software object models and database structure models, can use metrics.

Google created the Go programming language in 2007 to improve development productivity and run-time performance. It is a compiled, concurrent, garbage-collected, statically typed open-source programming language. The designer of Go wanted to address the weakness of other programming languages [5]. Go was released first in 2012 [6]. The current version is 1.15 that was released on 11 August 2020. Go is being widely used in Google to build high-performance network server [6]. Go is not a class-based object-oriented language like C++ and Java. But a OOP-style programming that can be implemented via the mechanism of interfaces [7]. The popularity of the language is increasing. According to Stackoverflow developer survey, 2020 Go is the third most wanted programming language [8].

Since the early 1990s, object-oriented (OO) techniques have been common in software development. Researchers have proposed several metrics, including metrics for Chidamber and Kemerer (CK) [9] and object-orientated design (MOOD) [10] for quality assurance of OO software.

This paper presents an empirical analysis of GoLang projects to determine how they developed. It also studied whether GoLang projects propagate any code smells. This study considered distinct Go projects from different software domains, e.g. library, web service, etc. to confirm the uneasiness findings. This paper concludes that almost every GO project consists of code smells especially feature envy which is one of the critical security issues in source code. This paper also depicted that developers rarely used deep inheritance in GO project, which might increase the chance of duplicate code.

2 Related Work

GO is one of the rapidly grown popular languages in the software developer community, which is significantly different from other languages in terms of syntax and other features. B. Shaikh et al. demonstrated the implementation differences between the other programming languages with regards to GO [11]. R. Subramanyam et al. provided empirical evidence supporting the role of CK metrics in determining software defects and they found that these metrics on defects vary across the samples from two programming languages—C++ and Java [12]. Yasir et al. built GodExpo, a tool that can detect God Structures in GoLang programs using metrics Weighted Method Count, Tight Class Cohesion, and Access to Foreign Data [13]. Several software metrics were also presented to analyse object-oriented systems [14] where 10 traditional metrics such as Line of Code (LOC) and CK metrics were used for the experiment. However, their work experimented on 3 Python and 5 Java projects.

An empirical exploration of Quality Measurement for JavaScript solutions has been measured with a list of software quality metrics [15].

Rosenberg et al. assessed Object-oriented Assurance of Quality, and Object-oriented Risk Assessment [16]. Aggarwal et al. proposed a series of metrics related to different buildings such as class, coupling, cohesion, hiding information, polymorphism, and reusability [17]. Briand et al. provided the framework for Object-Oriented System cohesion and coupling measurements [4].

F.B Abreu presents a MOOD array of paradigms that defines the use in software code of object-oriented paradigms [18]. These measures contribute to an object-based system assessment of efficiency and productivity. Software development requires a measurement mechanism for feedback and evaluation. Without measurement of software process, project planning becomes difficult. Victor R. Basili et al. describe Goal Question Metric (GQM) approach to measure and evaluate software process [19].

MOOD is referred to as encapsulation (MHF, AHF) as the basic structural mechanism of the object-oriented paradigm [20], inheritance (MIF, AIF), [21] polymorphism (POF), and message passing (COF). Two main features are used in MOOD models: methods and attributes in every metric. Methods are used to perform many types of operations on objects such as achieving a status change. Attributes are used to represent the status of each entity in the system. The study Neelamegam, C. and D. Punithavalli focus on a set of object-oriented metrics that can be used to measure the quality of an object-oriented design [22].

In the field of fault forecasting, several researchers have done significant work. The literature survey consists of developing CK parameters to explore various techniques used to model fault prediction. For object-based (OO) code, CK metric suit is the most used metrics. Chidamber et al. [23] have developed and implemented a new set of Object-Oriented Design application metrics. They found that Object-Oriented can contain some of the application crisis solutions.

Although the several empirical studies were performed by MOOD and CK metrics on varieties of programming languages, these types of studies are limited for GO projects.

3 Dataset Description

Five open-source GitHub projects have been experimented on for evaluating the tool's utility. The default threshold has been used while running the projects using the tool. The five real-world projects are selected from different categories to avoid biasness in the data set, which are listed below.

1. **Hugo:** It is a static HTML and CSS website generator.[1]
2. **Protobuf:** This project implements GO binding for protocol buffers.[2]
3. **OAuth2:** This is online authentication system version 2 implementation.[3]
4. **Groupcache:** Groupcache is a caching and cache-filling library, intended as a replacement for Memcached in many cases.[4]
5. **Mock:** It is mocking framework for GO programming language.[5]

4 Methodology

This paper uses eight well-known software metrics equally from MOOD and CK to analyse the internal architecture of GO projects. It also found the GO projects' correctness index using two standard code smells named as god class and feature envy. To empirically analyse them, at first the targeted dataset was collected from GitHub and then processed for measuring the code. However, the code is statistically analysed. Then the projects were tokenized to calculate the value for god class and feature envy detection. The code metrics and code smells that are used in this study are explained below.

4.1 Line of Code (LOC)

The express "lines of code" (LOC) is a metric by and large utilized to assess a software program or codebase concurring to its estimate. It could be a common identifier taken by including up the number of lines of code utilized to type in a program.

4.2 Total Comments

In computer programming, a comment may be a programmer-readable clarification or explanation within the source code of a computer program. They are included with the reason of making the source code less demanding for people to get it, and are by and large overlooked by compilers and translators.

[1]https://github.com/gohugoio/hugo, "Hugo," retrieved 30 October 2020.
[2]https://github.com/golang/protobuf, "ProtoBuf," retrieved 30 October 2020.
[3]https://github.com/golang/oauth2, "OAuth2," retrieved 30 October 2020.
[4]https://github.com/golang/groupcache, "groupcache," 30 October2020.
[5]https://github.com/golang/mock, "Mock," retrieved 30 October 2020.

4.3 Weighted Method Count (WMC)

The Weighted Method Count or Weighted Method per Class metric was originally defined in A Metrics Suite for Object-Oriented Design [9]. The WMC metric is characterized as the sum of complexities of all methods declared in a class. This metric is a great indicator of how much exertion will be vital to preserve and create a particular class.

4.4 Lack of Cohesion of Methods (LCOM)

Lack of cohesion in methods. Cohesion refers to the degree of the intra-relationship between the elements in a software module such as packages and classes [9]. Ideally, each element has a strong relationship in the module by achieving a particular functionality. The LCOM metric indicates a set of methods in a class is not strongly connected to other methods.

$$\mathtt{LCOM} = \frac{(m * (a/V))}{m - 1} \tag{1}$$

Where

I. m: the number of methods in the class,
II. a: the number of methods in a class that access an instance variable,
III. V: the number of instance variables.

4.5 Attribute Hiding Factor (AHF)

This is the degree of the invisibility of attributes in classes. The invisibility of an attribute refers to the rate of the overall classes from which the attribute is invisible. It can be calculated by summing the invisibility of each property with regard to the other classes within the project [10]. Within the previous calculation of invisibility, private = 1, public = 0, protected = Size of the legacy tree/number of classes. AHF is a fraction.

$$\mathtt{AHF} = \frac{\sum_{i=1}^{TC} A_h(C_i)}{\sum_{i=1}^{TC} A_d(C_i)} \tag{2}$$

Where

$A_h(Ci)$:hidden Attributes in class C_i,
$A_d(Ci) = A_v(C_i) + A_h(C_i)$:Attributes defined in C_i,

$A_v(Ci)$:visible Attributes in class C_i,
TC :Total number of Classes.

4.6 Method Hiding Factor (MHF)

MHF is the measure of the invisibility of methods in classes. The invisibility of a method refers to the percentage of the total classes from which the method is invisible. It can be calculated by summing the invisibility of each method with respect to the other classes in the project [10]. In the previous calculation, private = 1, public = 0, protected = size of the inheritance tree/number of classes. MHF is a fraction.

$$\texttt{MHF} = \frac{\sum_{i=1}^{TC} M_h(C_i)}{\sum_{i=1}^{TC} M_d(C_i)} \tag{3}$$

Where

$M_h(Ci)$: hidden Methods in class C_i,
$M_d(Ci) = M_v(C_i) + M_h(C_i)$: Methods defined in C_i,
$M_v(Ci)$: visible Methods in class C_i,
TC: Total number of Classes..

4.7 Attribute Inheritance Factor (AHF)

This is the fraction of class attributes that are inherited. The expression for calculating it requires summing the inherited attributes for all classes from its super-classes in a project [10].

$$\texttt{AIF} = \frac{\sum_{i=1}^{TC} A_i(C_i)}{\sum_{i=1}^{TC} A_a(C_i)} \tag{4}$$

Where

A_i : inherited Attributes,
$A_a(Ci) : A_d(C_i) + A_i(C_i)$,
A_d: defined Attributes,
TC: Total number of Classes.

4.8 Method Inheritance Factor (MHF)

This is obtained by dividing the total number of inherited methods by the total number of methods [10]. The total number of inherited methods is obtained by the sum of the number of operations that a class has inherited from its super-classes.

$$\mathtt{MIF} = \frac{\sum_{i=1}^{TC} M_i(C_i)}{\sum_{i=1}^{TC} M_a(C_i)} \tag{5}$$

Where

$M_a(C_i)$: $M_d(C_i) + M_i(C_i)$,
M_i:inherited Methods,
M_d:defined Methods,
$\boldsymbol{TC}$:Total number of Classes.

4.9 Code Smell for GO Lang

Traditional software development processes emphasise on maintainability of software. The Good practice is to improve maintainability by detecting bad smells and refactoring them.

We measured 2 code smells from each project. The 2 code smells are i. God Struct and ii. Feature envy God Struct in GoLang is similar to God class in C++, Java. This denotes a Struct with extensively large code and has a harmful impact on maintainability [24]. Feature envy denotes a method or function that uses the data of another object more than its own data. It hampers the code organization [25].

5 Analysis and Discussion

Table 1 presents information regarding CK metrics of the five projects. The first column shows the project name and next four columns represents Lines of Code (LOC), Comments, Weighted Method Count (WMC), Lack of Cohesion of Methods (LCOM).

Table 2 presents information regarding four mood metrics of projects Attribute Hiding Factor (AHF), Method Hiding Factor (MHF), Attribute Inheritance Factor (AIF) and Method Inheritance Factor (MIF). The more value of AHF and MHF refers to having more encapsulation in projects. The less value of AIF and MIF refers to less dependency on the hierarchy level. The table shows that projects have significantly lower values of MHF, AIF and MIF.

Table 1 CK metrics of the projects

	LOC	Comments	WMC	LCOM
oauth2	7851	678	765	7
groupcache	7636	492	1127	31
hugo	173958	8551	16268	339
protobuf	601264	17568	115288	999
mock	16743	1574	2154	109

Table 2 A metrics results of the study

	AHF	MHF	AIF	MIF
oauth2	0.317073	**0.042017**	**0.046512**	**0.012448**
groupcache	0.5	**0.042672**	**0**	**0**
hugo	0.580942	**0.111334**	**0.087**	**0.015573**
protobuf	0.157031	**0.003518**	**0.053378**	**0.001777**
mock	0.75	**0.032599**	**0**	**0**

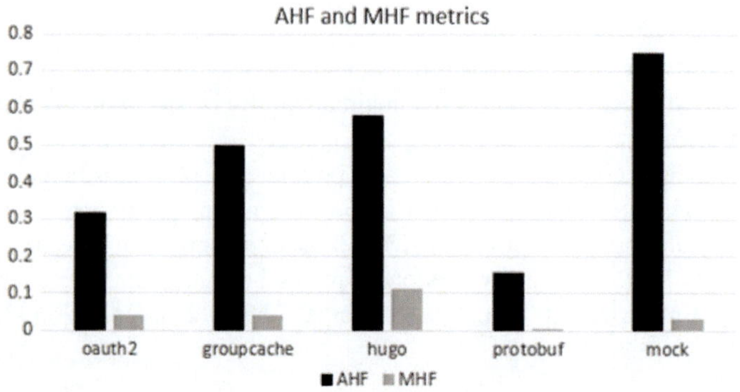

Fig. 1 AHF and MHF of the projects

Figure 1 shows the graphical representation of measured AHF and MHF of the projects in Bar diagram. Here, it is easily understood that the MHF of the projects is very low. The maximum MHF is found in huge projects with a value of 0.111334.

Figure 2 shows a graphical representation of AIF and MIF in a bar diagram. The results show both AIF and MIF having lower values. The maximum value of AIF is 0.053378 and the minimum with 0. The maximum value of MHF is 0.015573 and the minimum with 0. Mock projects have a value of 0 for both AIF and MIF.

Table 3 presents information regarding the total code smell of these five projects. The first column represents the project name and the next two columns represent the total God Struct and total feature envy of the projects.

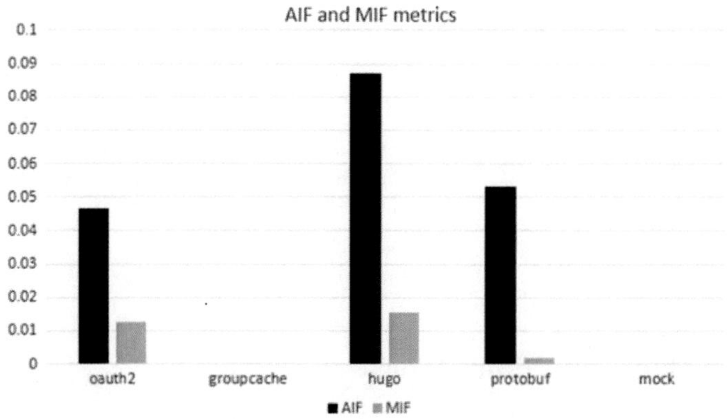

Fig. 2 AIF and MIF of the projects

Table 3 Code smell of the projects

	Total god struct	Total feature envy
hugo	17	27
protobuf	9	10
oauth2	0	12
groupcache	0	1
mockmaster	1	0

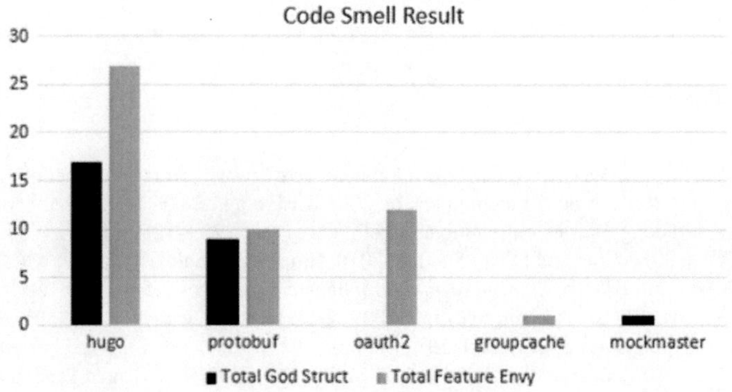

Fig. 3 Code smell

Figure 3 represents the total God Struct and total feature envy of the projects. God Struct value ranges from 0 to 17 and feature envy from 0 to 27. There less code smell is detected for the projects.

5.1 Findings and Recommendations

The statistics of the result are significant for MOOD metrics and the measured code smells. The three MOOD metrics MHF, AIF, MIF values are significantly low through the projects. It denotes that Object-Oriented paradigm is loosely followed in the projects. On the other hand, the analysis of the two code smells shows the projects having lower values that signify few smells.

6 Conclusion and Future Work

The study presents CK metrics, MOOD metrics and code smells within projects written in GoLang. The total God Struct and total feature envy of the projects are measured. The projects that have been experimented on are open-source and can be found in GitHub. The results give an overview of the metrics that can be followed by the developers to choose GoLang as an appropriate programming language for their software.

The world of programming is moving continuously. New changes come in with the time and problems which exist at present will be solved in future. Therefore, Future work will be performed with new the context of GO programming and with more dataset for result accuracy for building automation tool to measure the quality of GoLang source code and to detect fields of improvement.

References

1. Sheta A, Rine D, Ayesh A (2008) Development of software effort and schedule estimation models using soft computing techniques. In: 2008 IEEE congress on evolutionary computation (IEEE world congress on computational intelligence) (pp. 1283–1289). IEEE
2. Kaur, A., K.S. Kahlon, and D.P.S. Sandhu. 2010. Empirical Analysis of CK & MOOD Metric Suit. *International Journal of Innovation, Management and Technology* 1: 447.
3. Barkmann, H., R. Lincke, and W. Low. 2009. Quantitative Evaluation of Software Quality Metrics in Open-Source Projects. IEEE.
4. Basili, V., L. Briand, and W. Melo. 1996. A Validation of Object Oriented Design as Quality Indicators. *IEEE TSE* 22: 751–761.
5. Pike, Rob. 2012. *Go atGoogle: Language Design in the Service of Software Engineering.* Retrieved from https://talks.golang.org/2012/splash.article [accessed: 30 October,2020].
6. The Go Project. 2012. https://golang.org/project [accessed: 30 October, 2020]
7. Todorova, M., M. Nisheva-Pavlova, G. Penchev, T. Trifonov, P. Armyanov, and, A. Semerdzhiev. The Go Programming Language: Characteristics and Capabilities. Annual of "Informatics" Section Union of Scientists in Bulgaria, 6: 76–85.
8. 2020 Developer Survey. 2020. *In February 2020 nearly 65,000 developers told us how they learn and level up, which tools they're using, and what they want.* Retrieved from https://insights.stackoverflow.com/survey/2020#technology-most-loved-dreaded-and-wanted-languages-loved [accessed: 30 October, 2020]

9. Chidamber, S., and C. Kemerer. 1994. A Metrics Suite for ObjectOriented Design. *IEEE Trans* 20: 479–493.
10. Harrison, R., S. Counsell, and R. Nithi. 1998. An Evaluation of the MOOD Set of Object-Oriented Software Metrics. *IEE* 24: 491–496.
11. Shaikh, Mrs. Bareen, and, Sangeeta Borde. 2013. Quantitative Evaluation by applying metrics to existing "GO" with other programming language.
12. Subramanyam, R., and M.S. Krishnan. 2003. Empirical analysis of CK metrics for object-oriented design complexity: implications for software defects. *IEEE Transactions on Software Engineering* 29 (4): 297–310.
13. Yasir, R.M., M. Asad, A.H. Galib, K.K. Ganguly, and, M.S. Siddik. 2019. GodExpo: an automated god structure detection tool for Golang. In *2019 IEEE/ACM 3rd International Workshop on Refactoring (IWoR)* (pp. 47–50). IEEE.
14. Destefanis, G., S. Counsell, G. Concas, and R. Tonelli. 2014. Software Metrics in Agile Software: An Empirical Study. In *Agile Processes in Software Engineering and Extreme Programming. XP 2014*, vol. 179, ed. G. Cantone, and M. Marchesi, Lecture Notes in Business Information Processing. Springer.
15. Kostanjevec, D., M. Pusnik, M. Hericko, B. Sumak, G. Rakic, Z. Budimac. 2017. A Preliminary Empirical Exploration of Quality Measurement for JavaScript Solutions. In *6th Workshop of Software Quality, Analysis, Monitoring, Improvement, and Applications*, Belgrade, Serbia.
16. Rosenberg, L.H., and L.E. Hyatt. 1997. Software quality metrics for object-oriented environments. *Crosstalk Journal* 10 (4): 1–6.
17. Kaur, Gurpreet, and Mehak Aggarwal. 2014. A Fuzzy Model for Evaluating the Maintainability of Object Oriented System Using MOOD Metrics.
18. Abreu, F.B., R. Esteves, and M. Goulão. 1996. July. The design of eiffel programs: Quantitative evaluation using the mood metrics. In *Proceedings of TOOLS'96*.
19. Caldiera, V.R.B.G., and H.D. Rombach. 1994. The goal question metric approach. *Encyclopedia of software engineering*, pp. 528–532.
20. Parnas, D. 1972. On the Criteria to be Used in Decomposing Systems into Modules. *ACM* 15: 1053–1058.
21. Kitchen, B., N. Fenton, and L. Pfleeger. 1995. Towards a Framework for Software Measurement Validation. *IEE* 21: 929–944.
22. Neelamegam, C., and D. Punithavalli. 2010. P a g e l 183 Global Journal of Computer Science and Technology A Survey- Object Oriented Quality Metrics.
23. Chidamber, Shyam R., and Chris F. Kemerer. 1994. A Metrics Suite for Object Oriented Design, IEEE, vol. 20.
24. Pérez-Castillo, R., and M. Piattini. 2014. Analyzing the harmful effect of god class refactoring on power consumption. *IEEE software* 31 (3): 48–54.
25. Guru, Refactoring. 2014. *Feature Envy, Signs and Symptoms*. Retrieved from https://refactoring.guru/smells/feature-envy [accessed: 31 October, 2020].

Author Index

S. Anwar and A. Rauf (eds.), *Proceedings of the First International Workshop on Intelligent Software Automation*, Advances in Intelligent Systems and Computing 1347, https://doi.org/10.1007/978-981-16-1045-5